OFFENSE
OF
REASON

DISCERNING TRUTH FROM
DISSEMBLING NARRATIVES

———

MAURIZIO
DiMAURO

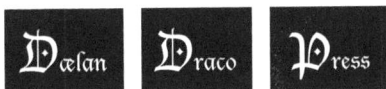

✝

Dælan Draco Press

First published by Daelan Draco Press 2020

Copyright © 2020 Maurizio DiMauro

First edition

ISBN: 978-1-7357568-3-7

Cover art by Charlotte Daniels

For Emmett

In absentia lucis, Tenebrae vincunt.

Table of Contents

Prologue

We learned in school that people in ancient times were led to believe in demons, saints, anthropomorphized gods, as allegorical devices for instilling ethics, morality, virtue, practicality, and for imposing variously just or unjust controls over populations.

The widely held view today is that, thanks to science, we have surpassed such indoctrinations. Climate science, not religion, supposedly persuades us to take measures to "save the planet". Medical science, not superstition, ostensibly tells us that vaccines are safe, effective and absolutely a necessary part of normal life.

That science should be the source of knowledge, rather than religion or superstition, is a reasonable proposition. But what if you don't know the true meaning of *scientific method*? What if you wouldn't be able to tell the difference between an experiment conducted by the scientific method and one that wasn't? How are you to judge, for example, the soundness of approaches, conclusions and prescriptions propounded by various authorities? How are you to gauge whether what you're told is in fact a belief akin to religious dogma merely disguised as science?

As I describe in the ensuing pages, starting with the modern global climate narrative in the first chapter, many government-backed mainstream beliefs can profoundly alter people's lives and even curtail human rights. Taking a government-endorsed position at face value or delegating one's critical thinking faculties to its proponents is highly inadvisable. The main purpose of this book is to avail you with a greater ability to make your own reasoned conclusions as to the validity of any big claim, irrespective of its popularity or heterodoxy and largely independent of your particular expertise.

To that end, I closely examine familiar examples of prominent theories in this light, several which are commonly accepted as true and one generally regarded as false but which enjoys something of an influential cult following in alternative circles (Chapter VI).

Over the course of that exposition I develop an overall theory-evaluation methodology, culminating in an organized set of practical criteria presented in Chapter V.

I. The Theory of Stuff Happens

A CLIMATE OF DOOM

Allow me to place a truism in the forefront of your mind: carbon dioxide is indisputably an essential molecule of life. Without it we wouldn't be possible. We animals breathe it out, and plants use it to make our food.

Authorities the world over claim that this very molecule is killing the planet, and we're to blame. Thus we must drastically curtail our production of this dangerous chemical, lest we reach an irreversible, runaway, catastrophic[1] global heating process. [1]

---

[1] There is no shortage of examples of doomsday assertions from the mainstream press. For instance, National Geographic Magazine (March 2020) [4], The Weather Channel (July 2020) [5], Washington Post (June 2020) [6] and The New York Times (Nov 2019) [7].

That reasoning forms the basis for policies designed to curtail our collective *carbon footprint* [3], including taxation and the establishment of *carbon markets*[2] through *emissions trading*[3] as variously punitive and financially incentivized energy rationing schemes for processes involving the release of carbon dioxide.

In the name of saving the planet we're implored to prepare for the coming of a radically reimagined way of life. Meat and meat products should be banned[4]; petrol cars should be phased out entirely to help meet the warming cap of 1.5 degrees Celsius above *pre-industrial levels* set by the Paris Accord[5]; we must practically give up the luxury of air travel [12]. Unless we stop polluting[6] the Earth with carbon dioxide[7], the narrative asserts, we're all doomed.

---

[2] Article 6 of the Paris Agreement [8].

[3] Article 17 of the Kyoto Protocol [9].

[4] [10][11].

[5] Article 2 of the Paris Agreement [8].

[6] [2][13][14][15].

[7] Note that the United States Supreme Court in *Massachusetts v. EPA* explicitly adopts the view that carbon dioxide is a pollutant. See page 4 of the syllabus [13].

The message long and broadly promulgated in the mainstream press is that the science is settled and that Climate Change[8] dissenters are anti-science. Wide effort is made to silence[9], discredit or deride[10] scrutiny and independent review of the establishment climate credo.

To summarize the Climate Change ethos:

1) CO_2 qualifies as a literal pollutant;

2) Open debate is systematically suppressed or preempted;

3) Contrarian views that reach the surface are derided; and

4) A near-cultish doomsday obsession permeates common discourse.

---

[8] Also known as Anthropogenic Global Warming.

[9] For example, the BBC in 2007 reversed its erstwhile "equal weight" policy toward the subject of climate, stating that, "The BBC... has come to the view that the weight of evidence no longer justifies equal space being given to the opponents of the consensus," [16][17]. More recently, in 2018, the BBC has bolstered that stance by issuing formal internal guidance to its journalists, intimating that, "The BBC accepts that the best science on the issue is the IPCC's position," [18].

[10] For example, author Michael J. I. Brown in a 2015 PHYS ORG article [19] asserts that *peer-reviewed* works in disagreement with the alleged "scientific consensus" have the effect of maintaining an "illusion of doubt and uncertainty" around Climate Change.

Extraordinary claims, and accompanying countermeasures, especially those entailing forced and drastic personal life changes, require unambiguous evidence and grave scrutiny. But how are we to judge the veracity of Climate Change orthodoxy if we're mere mortals with no advanced degrees in the specialized field of climatology? It would seem that we're stuck blindly following the establishment on this topic.

Although Earth's climate is no doubt a complex subject on which very few of us can expound credibly, it turns out that the central claims of Climate Change can be examined without being a professional climatologist. In fact it helps if you aren't, so that you have fewer preconceived notions[11] at the outset.

We have some pertinent questions to ask. What exactly is the theory of Climate Change, and should we believe it? Is it worthy of consideration? Is it valid? That requires a definition of "valid." To that end I will in the next section turn to the philosophy of falsifiability.

---

[11] Bear in mind also that those who've built successful careers on the basis of Global Warming have a vested interest, consciously or not, in it being true. Therefore outsiders have an objectivity advantage.

FALSIFIABILITY FOUNDATIONS

A falsifiable theory, putatively first propounded by Karl Popper in the early 20th century [20(a-c)], is one which admits tests that could prove it false. It's also the idea that one can never strictly prove a theory true.

The falsifiability principle is foundational to modern scientific thought [152], which includes any claim based on reason, logic and evidence. That implies that any theory purporting to be scientific in this sense must pass the falsifiability test to be valid. Any such claim must be able to pass its own authenticity test.

In this book I will develop something of a falsifiability practical guide by way of examples of allegedly scientific claims which variously turn out to be either falsifiable or unfalsifiable. For claims which turn out to be falsifiable, I follow up by showing how some very established theories are provable as false by virtually anyone with common sense.

WHICH TYPE OF UNIVERSE DO WE INHABIT?

You may wonder why falsifiability is asymmetric. Why does it hold that we may only prove theories false? One reason is as follows.

Consider a universe governed by discoverable natural laws. Let's call this the *alpha* world. In alpha world we can never presume to completely understand the natural laws. We can only assume our models of them are approximate. This is because our limited domain of spatiotemporal existence in such a world prevents sampling all there is to see of it, even though we see and comprehend more and more of it as time passes. What we can predict in alpha world is always limited to some context, however expansive and expanding it may be.

In popular science culture, such a limitation is essentially embodied in the "streetlight effect", which I'll paraphrase as the theory that the keys your spouse allegedly dropped outdoors don't exist, because you cannot see them on the ground (under the streetlight). As far as you're concerned, your open-ended theory is correct and will always be correct.

However, if you assume that you live in an alpha world, where there may be relevant information or variables to which you are not privy yet (unknown unknowns), you conclude that evidence supporting your theory doesn't constitute proof that it will *always* predict correctly. You reason that it is not possible to prove that the keys *don't* exist, because you don't know that you're seeing every part of the universe where the keys could be; you may see only what you *can* see at the moment.

As in the streetlight effect, falsifiability assumes we live in alpha world; and in that world, as you can see, there can be no absolute certainty about the future state of knowledge, only knowledge about the past. When you find the keys, after your spouse brings a flashlight illuminating regions away from the street lamp, the new variable s/he illuminated—no pun intended—has enabled you to prove your theory wrong.

Such is the limitation of theories in alpha world; they necessarily admit no tests which could prove them true.

In apparent contradiction to the latter statement, it is of course possible to prove true a finite set of predictions if they come true. The claim that stock X will rise by Y date and then fall by Z date admits tests of both veracity and falsity. It is in fact symmetric in the sense above. We just wait for the appropriate times to pass and observe what happens.

But the latter type of claim is either an instance of an implied theory making open-ended predictions about the future state of knowledge (a method claiming to predict price changes, in this case) or a random guess (untestable model). If it's an instance of an open-ended theory, like all such theories in alpha world, it cannot be proved true. If it's a random guess, it by definition admits no test, neither of veracity nor of falsity.

Now consider the opposite universe, *beta* world, one not governed by laws. In other words, there are no inherent, effectively consistent rules to discern[12]. Therefore, in beta world nothing is predictable, no matter how hard we try. In such a world we wouldn't be able to place satellites in orbit, because gravity wouldn't exist, at least not consistently. We couldn't conceive of any transportation mode, because there would be no rule governing motion; we couldn't develop the practice of farming, be able to learn how to hit a baseball with a bat, anticipate the right time to ask the girl of your dreams out to dinner... you get the idea. The assumption we're evidently forced into is that we live in alpha world, where falsifiability reigns or appears self-evident.

---

[12] This is equivalent to the rules changing unpredictably.

THE SCIENTIFIC METHOD

If I claim that all objects fall to the ground if I release them from a height at rest and that they all accelerate in the same particular way (absent air resistance), I'm claiming that *all* objects will *always* fall that way under those conditions. This is the general notion behind the familiar classical gravity model.

Let's parse this carefully: Others who successfully repeat the experiment themselves could only conclude that 1) those specific objects tested—i.e., not all objects in the universe—fell in identical ways and 2) those objects exhibited such behavior in the *past*. The results can prove nothing about the future nor about how objects not part of the experiments behave.

In a similar vein as the streetlight effect, we could very well discover that a previously untested object, from a previously uncharted corner of the galaxy, does not fall the way all the others have in previous experiments. This novel object may not even fall at all; it may instead rise. We have no logical basis for discounting the latter possibility.

In this way, falsifiability is the cornerstone of the scientific method, which is another way of describing the limit of what we can know for certain. So I will make wide, nuanced application of the principle in this book, particularly in the context of highly personally impactful claims made by governing authorities, professional establishments and otherwise influential or consequential institutions.

AN INTRODUCTION TO REAL WORLD FALSIFIABILITY

Suppose I made the following claim: All swans are white. The White Swan theory is falsifiable[13]; the moment we observe one that is not white is the moment we can unambiguously conclude that the theory is false. [20(a-c)]

Now suppose I claim that there is such a thing as an Invisible Pink Unicorn[14,15]. Is IPU theory falsifiable? The claim is that it is pink but that one can never detect what color it is, because the IPU is invisible by definition. Since there is no test that could prove IPU false, IPU is not falsifiable.

---

[13] [20c], p. 4.

[14] Contemporary analogue of Bertrand Russell's teapot analogy. "Russell's Teapot." In Wikipedia, July 25, 2020. https://en.wikipedia.org/w/index.php?title=Russell%27s_teapot&oldid=969461605

[15] "Invisible Pink Unicorn." In Wikipedia, March 11, 2020. https://en.wikipedia.org/w/index.php?title=Invisible_Pink_Unicorn&oldid=945054897

But those are textbook examples. In practice, things get a little messy. A theory may seem falsifiable in principle but may not be falsifiable in practice, or it may be composed implicitly of multiple theories in the form of hidden assumptions or prejudices, each of which would have to be identified and examined under the lens of falsifiability.

Imagine that the government announced the existence of a class of deadly, invisible particles, called *units,* which maim and sometimes kill people by directing their bodies to make an uncontrollable and damaging number of unit copies. The two hundred-year-old theory further holds that units spread from host to host.

Unitalogists and mainstream scientists unquestionably believe units to exist and to comprise an unknown number of types, causing known and yet to be discovered illnesses.

One day the government announces the suspected emergence of a novel unit, dubbed Unit Recherché (UR). Because of UR's hypothesized lethality, the government requires all people to submit to personal isolation and occupational leave of absence, exacting profound psychological and economic tolls on the populace. The government justifies the deleterious measures citing the better-safe-than-sorry ethos.

That's serious. We ought to examine the claims carefully before we decide to obediently comply with such Draconian impositions, lest they devolve us into totalitarianism, for example. We must therefore parse the validity of UR, but ideally that of unital theory as a whole.

It would seem that proving that units don't exist in the first place, if true, would in principle do the trick to discredit the theory. But how could anyone prove the non-existence of something? Look! There it isn't! There it isn't again? It would require being able to account for all regions of the known and unknown universe. I'm afraid it's not possible[16].

So let's take a different tack. Perhaps it's possible to prove that UR, or any other alleged type of unit, is not found in the presence of illness it presumably causes, implying an alternate cause. We might get somewhere now.

In this hypothetical example, suppose that no experiments have ever demonstrated an association, let alone causation, between UR and its supposed illness. In that case it would seem that we'd be done. Unital theory has no support.

---

[16] See Popper [20c], pp. 48-50, for a rigorous treatise on existential versus non-existential statements regarding falsifiability.

But hang on, that doesn't imply anything about the overall theory of units. Just because UR, which is just one type of unit, doesn't cause illness doesn't mean a different type conceivably couldn't. Even if we tested several types and still found no illness associations we couldn't conclude unital theory false, because the theory posits the existence of both known and unknown units causing disease. So it would seem that it's not possible to ever prove unital theory wrong, even if it were wrong.

Would you be okay with following personally consequential government orders based on a theory no one could ever possibly prove wrong even if it were wrong?

But shame on us, in this hypothetical example, for having accepted the theory as true for two centuries when it was not falsifiable in the first place. So an important lesson here is to know that when anyone makes a claim it's his or her burden to define it in a way that could be proved false if it were false[17].

Notice that we didn't have to know anything about the faculty of unitalogy to determine whether it is scientific

---

[17] This example is inspired by the germ theory of viruses. How do you know viruses exist and cause disease, for example? You now at least have the means to examine that theory for falsifiability. You can find more insight on this topic in Chapter II.

(falsifiable). As I will develop in this book, we can determine falsifiability having no professional expertise ourselves in the applicable subject, provided we have access to constitutional definitions. If the theory happens to be falsifiable, we may need to cultivate some expertise before attempting to falsify it, but we generally need not be trained experts or specialists in the field.

In that spirit, let's turn to the theory of Climate Change.

A GREEN HOUSE OF CARDS

Despite the theory's name suggesting something not much more consequential than "Stuff Happens", Climate Change theory means something a little more specific:

The Earth is warming too much from the Greenhouse Gas Effect (GHE) due to an excess in anthropogenic Greenhouse Gasses (GHGs), principally CO_2. [1][21(a-d)][22]

Climate Change used to be called Anthropogenic Global Warming (AGW) theory, for obvious—thankfully more descriptive—reasons, but I will use both terms interchangeably.

Taken in its entirety, it's not immediately clear whether or not the AGW principle is falsifiable, because the theory has some potential ambiguities. For example, what does "excess" mean? It presupposes knowledge of what constitutes the right concentration of GHGs above which the atmosphere is presumably in excess. We would have to examine theories, if they exist, claiming to correctly establish the normal, acceptable or healthy concentration ranges for each GHG individually.

Similarly, AGW theory also presupposes that we've somehow been in the right global temperature and therefore that any increase in temperature would be a bad thing, which further begs the question of, what is too much additional warming or what is the right global temperature? Again, theories claiming to know the right global temperature, if they exist, would themselves have to be examined as to falsifiability.

Therefore Climate Change comprises multiple interdependent theories. To reduce Climate Change to something our falsifiability lenses can examine, what we really need to do is look for the part of the theory upon which every other implicit theory within it depends. And it would appear that we're in luck, because it turns out that the theory of Climate Change hinges on a critical hypothesis that is much more manageable than the theory as a whole.

It's called the Greenhouse Effect (GHE). Without it there can be no GHG-caused AGW, or no "Climate Change". It is literally the very mechanism by which any warming is said to happen above and beyond what would occur without GHGs. Therefore, 1) proving GHE theory false would logically render Climate Change itself false; and therefore 2) we would not have to bother trying to elucidate the aforementioned sub-theories or hidden assumptions of Climate Change nor examine the falsifiability of each. So let's focus our attentions on the GHE.

THE (RADIATIVE) GREENHOUSE EFFECT

What is GHE theory and is it falsifiable? For the GHE to work, all the following steps must occur [21(a-d)]:

1. The sun's rays hit the Earth's surface, raising its temperature[18].

2. That heated surface then radiates energy upward in a band of wavelengths that GHGs in the atmosphere can absorb.

---

[18] The GHE model assumes that sunlight does not appreciably increase the temperature of atmospheric air molecules. Therefore it assumes that sunlight essentially bypasses the air.

3. The temperature of the GHGs consequently increases, resulting in re-radiation of part of that energy back down to the surface (the rest out into space).

4. The addition of this "back-radiation" intensity from the GHGs to that from Earth's surface causes the temperature at the surface to be higher than it would be without (or with less) GHGs in the atmosphere.

As you can see from step 4 of the definition, the GHE theory hinges on the belief that radiation intensities may be added together to deduce final temperature; or, put another way:

GHE assumes that *temperature is a function of heat radiation intensity*. Is this intensity dependency assumption falsifiable? Yes, an experiment showing that changing the heat source intensity does not result in a temperature change would prove the intensity assumption false. So let's try that.

It turns out that a physical experiment is unnecessary here. We may draw from common experience, which I motivate with some rhetorical questions, at least one of which should elicit a eureka moment:

- Consider two identical rocks at the same temperature; each therefore individually emanates the same heat radiation intensity. We put the two rocks together such that the intensity we experience from them is twice that of one. Do we measure a higher temperature with two rocks than we do with one?

- Knowing that two identical candles have twice the intensity of one, is the temperature from two such candles higher than that from one alone?

- Does the temperature of your gas burner increase when you increase the size of its flames?

GHE theory would imply "yes" to all the above questions, even though the intuitive answer is "no". You can easily verify the above experiments yourself with a thermometer, particularly the first one on the list.

Instead, temperature is about *color*. Again from common experience, we know that a blue flame from a bunsen burner or stove top is hotter than a yellow one from a candle, no matter how big (intense) the yellow flame is. The combined flame of two candles does not change the temperature, because the color is still yellow.

It's no coincidence that astrophysicists and astronomers determine the temperature of stars by measuring their color spectra[19], not by measuring their illumination intensities; otherwise a closer star would always be hotter than an identical star further away[20].

Even infrared thermometers don't measure the *intensity* of infrared light from an object[21]. They instead sense the frequency (color) of that light, because it is the color of light which determines the temperature it imparts on a material. Focusing light on a medium creates a momentary temperature gradient which infrared thermometers convert to a voltage difference. That voltage is translated into a numerical value (temperature) presented in a display. [24]

---

[19] Light is composed of electromagnetic waves in an overall spectrum of wavelengths or frequencies, some of which are in the visible region, but most of which are invisible, such as radio waves, x-rays, infrared, ultra violet.

[20] See basically any introductory college level textbook of astronomy or astrophysics, and see [25] for the concept of blackbody radiation on which spectroscopic temperature measurement is predicated. On the web, Prof. J. Brau of the University of Oregon has an example of spectroscopic temperature measurement for his Astronomy 122 course [23a]. Also, the online book in [23b] mentions the same principle in the Spectroscopy section.

[21] The phenomenon is known as *blackbody radiation*. It's an approximation of the observation that matter radiates characteristic spectra of light, which includes infrared, as a function of its temperature. See for example [25], pp. 443-452.

If you're satisfied with that common sense proof, we're done: the Greenhouse Effect is proved false, and therefore so is Global Warming theory ("Climate Change").[22]

But if that common sense view doesn't fully satisfy you, see *Appendix i* for a more rigorous treatment. Fortunately, it requires barely high school level math.

---

[22] It's important to point out that proving theory X false does not imply that there couldn't possibly exist a different theory Y successfully making the *same* prediction as X, such as a modified original theory X. What I have showed here is that Global Warming theory, as defined by IPCC and adopted by governments and the mainstream press, is false.

II. Black Death Lite

THE PREMISE

As I write this in the summer of 2020, most of the world is in the grips of a supposed pandemic. We've been told of the outbreak of a novel virus so dangerous that it justifies authoritarian measures resulting in worldwide economic suffering, forced social isolation and curtailed basic human freedoms.

Those of us not naturally inclined to accept the official word face trying to verify extremely urgent, consequential claims for which most of us have little to no appropriate expertise.

What is the central claim of this alleged pandemic? During the ostensible ramp up of the supposed outbreak in the January-March timeframe of 2020 we were expected to believe that a novel virus was poised to kill many millions around the globe if we didn't shut down normal life as we knew it, agree to submit to medical testing, so-called contact tracing and other privacy violations, comply with mask-wearing mandates, abide by decrees calling for closures of businesses arbitrarily deemed non-essential and layoffs of similarly non-essential workers, follow orders that restricted our movements and gatherings (public and private) and outdoor activities, and so on.

GASLIGHTING 101

Governments and media initially gave the impression that one's chances of survival were around 3% if infected with the alleged virus, based on announcements such as from the World Health Organization (WHO) [26] which crudely divided the number of fatalities by the number of *reported* cases. That seemed to imply that one's chances of survival were ostensibly 97% if infected. But that figure turned out to be misleading even assuming a real virus.

What officials should rather have reported was something like the *infection fatality rate* (IFR) or the *case fatality rate*

(CFR)[23]. Instead, on that misleading 3% basis, authorities placed essentially the entire population under "lockdown", which is jail terminology. It meant that the government dictated when and if people could leave home, under threats of fines and incarceration. The government's allowances for being outside the house were meager, and they came with restrictions of their own: no gathering in groups greater than an arbitrary number, no entering public spaces without ostensibly protective face coverings, no bathing in

---

[23] CFR is one's chance of dying of a disease if one has it. It is the total number of deaths ascribed to the disease divided by the total number of people who have it (cases). It's the fraction of those ill who ended up dying of the illness, which equals (number of deaths) / (number of cases), where "cases" means *symptomatic* individuals. CFR requires estimation. One can't divide the number of deaths of that disease by the number of people known to be afflicted, because one can't assume awareness of every instance of illness, whereas all cases of death are reported. So instead, the number of people with the disease at large must be estimated by extrapolating from a random sample of the population. For example, if 5 out of 20 cases of disease in a random population sample die of the disease, the CFR is (deaths/cases) = 5/20 = 1/4. But if we don't take a random sample the number of cases of which we're *aware* is generally smaller, say 10, which misleads us to believe there are twice the number of deaths from disease than there are at large (5/10 = 1/2 compared to 1/4). Similarly as to IFR, except that it is (number of deaths) / (symptomatic + asymptomatic count). Government officials and the press had initially reported fatality rates based on *known* symptomatic counts (cases), which is generally an inflated number.

the ocean on public beaches, and other prohibitions, such as arbitrary closures of public parks.

Then a review of 23 published scientific papers[24] came out in June 2020 concluding that the IFR was actually a full order of magnitude less than what governments and media had originally (falsely) implied by the aforementioned con- flation of measures as a pretext for lockdowns. That new estimate, which is very close to that from the U.S. Center for Disease Control and Prevention (CDC)[25] a month ear- lier [28], was now 0.25%, roughly on par putatively with the

---

[24] A mixture of published and preprint peer-review publications conducted by distinguished Stanford University professor John P.A. Ioannidis. Since he reviewed 23 studies in total, the aggregate sam- ple size was upwards of 12,000 subjects. [27]

[25] The CDC reported something related to IFR, called SCFR ("Symptomatic Case Fatality Ratio" = 0.4%), which is the same as CFR. They also reported their estimate of the fraction of infec- tions which are symptomatic ("Percent of infections that are asymptomatic" = 0.35). IFR is the number of deaths divided by the total number of infections (symptomatic + asymptomatic), where- as SCFR is the same except we omit asymptomatic people from the calculation. So if we let F be the number of fatalities and $A+S$ be the total number of cases of infection, we have that IFR $= F/(A+S)$ and that SCFR $= F/S$. Notice that IFR is smaller by a factor $S/(A+S)$; this is the fraction of infections that are symptomatic, which corresponds to 1 minus the CDC's "Percent of infections that are asymptomatic" (asymptomatic fraction). Therefore, to convert SCFR to IFR, we just multiply CDC's SCFR by the symp- tomatic fraction to get $(0.4\%) \times (1-0.35) = 0.26\%$.

seasonal flu and similarly impactful diseases. Thus the IFR alone can't logically form the basis for alarm.

If the virus had turned out to be much more infectious than ordinary diseases, we could possibly be justified for sounding the alarm[26]. However, mainstream mathematical models attempted for such a purpose were ultimately discredited as wildly inaccurate[27].

But that didn't change the US government's position at all. Authorities around the world in fact kept operating under the stated belief that we had an inordinately dangerous pandemic on our hands justifying human rights violations resulting from a continued Draconian state of declared emergency, with strangling small business restrictions, mask-wearing mandates[28], ridiculous, impractical and scientifically contradicted [29] six-foot separation social distancing rules and other impositions.

---

[26] IFR is a fractional population estimate. Suppose disease A has the same IFR as disease B but is so contagious that *everyone* becomes infected. That would imply that a lot more people die of A than B.

[27] See for example [42-44] and [47].

[28] In the CDC's own Emerging Infectious Diseases journal is a peer-reviewed article stating that there is no scientific support for mask effectiveness, based on a review of 14 randomized controlled trials carried out from 1946 to 2018 [30].

Without a reliable model of infectivity, the initial, false implication of a 97% survival rate was not logically any kind of dire emergency. And the now mathematically correct 99.75% (IFR = 0.25%) rate meant that any notion of emergency didn't pass the laugh test altogether.

The establishment's own epidemiological estimation amounted to negligible impact, even if we take everything at face value. We're talking about a narrative implying near-total survival with no conclusiveness as to infectiousness. If that's not the definition of non-issue I don't know what is.

Over the ensuing months the central narrative would remain the same, and become more extreme. Many states and counties in the United States continued to issue decrees requiring social distancing[29], crippling capacity reduction or outright banning of certain small business operations, face coverings in all public and business spaces, 14-day self-quarantine rules if traveling across state lines or national borders, and so on. These continued to be mandated under threats of fines, imprisonment, business license revocation, even in the voluminous presence of

---

[29] 6-foot physical interpersonal separation between people in the US and generally 1-2 meters elsewhere.

mainstream scientific evidence contradicting the logic of such measures[30].

UNBOUNDED ALARMISM

The over the top sense of panic is rather reminiscent in tenor to the "Climate Change" doomsday prediction I described in Chapter I. In both cases authorities claim to know for certain that something horrible will happen to the entire world and that inaction results in dire, irreversible and unacceptable damage. The premise, crucially, is that we don't have the luxury of waiting to see what happens, because it's better to be safe than sorry. Therefore, we must assume authorities are right, even if they might be wrong.

The better-safe-than-sorry-argument in this context is a logical fallacy. The reason is that it presupposes no significant cost of taking the alarmist view as compared to the critical view. Instead, cost is essentially only ascribed to taking the dissenter's view. It draws parallels with the "mush-

---

[30] Governors across the U.S. have continued to issue executive orders requiring all people, for example, to wear cloth masks in public despite the sheer volume of published, peer-reviewed mainstream articles over the last decade concluding that such masks have no measurable protective effect against viruses [31-37].

room cloud" argument[31] delivered by the US National Security Adviser Condoleezza Rice in the drumbeat to the second invasion of Iraq in 2003 under the George Bush administration. We were told that Iraq undoubtedly had weapons of mass destruction justifying the ensuing death and mayhem in the name of defense against the alleged Iraqi threat.

The rationale for the Iraq invasion turned out to be false—flat out wrong or at worst an orchestrated deception aided by an uncritical, unquestioning mainstream media.

We know we were massively misinformed by all major news organizations and government. So it boggles the mind that the media should today be largely heeded as though it couldn't possibly lead people astray on monumentally consequential topics again. It's reason alone, at a minimum, to examine the premises and costs associated with taking government or media pronouncements at face value.

The economic impact alone of executing pandemic countermeasures to an alleged pandemic is a widely acknowledged cost that led to bankruptcies and otherwise

---

[31] "But we don't want the smoking gun to be a mushroom cloud." [38]

personal financial ruin. As of this writing, the number of people who are consequently unemployed, by official count, has rivaled Great Depression era levels in the U.S. [39-41].

There are also health costs associated with restrictions imposed on personal lives, such as suppressed physical health stemming from reduced sunshine exposure and degradation of mental health resulting from extended social isolation.

More crucially, history warns that relinquishing human rights per se ultimately exacts the highest cost, because it paves the way for government-sponsored atrocities: Stalin, Hitler, Mao Zedong, to name but a few responsible for unfathomable scales of outright murder. On all those accounts the authoritarian regimes imposed tyranny in the guise of bettering the human condition in some way, the same essential pretext of the 2020 pandemic narrative. So at the very least, it behooves us to analyze carefully the narrative we're given.

ANATOMY OF A PANDEMIC

As a technical claim, the pandemic alarm is highly loaded with ostensibly scientific premises and assumptions seemingly inscrutable to the layperson. Though being able to examine the scientific claims themselves would be ideal, I will show that it's not strictly necessary.

Imagine that our entire civilization consists of just one thousand people living on a small island, the only land on the planet. We live quite well, each of us with respective roles in commerce and daily life. We're a free society and we depend on mutual contributions for survival, happiness and well-being.

It's winter and we notice the usual number of people under the weather, and a few elderly folks die, as happens every year. But this time the island's influential and well regarded elders claim that the cause of those illnesses and deaths is a novel communicable disease caused by an invisible agent affecting everyone. They claim a person can have the disease and infect others without themselves feeling ill.

Without offering a basis for their claims, they announce that chances of dying from it are around 3% of those infected. That doesn't seem very high to most islanders, but the elders swear that the disease is so contagious that soon everyone will have it. They base this on a model having admittedly high uncertainty due to the paucity of data available at this stage and due to the nature of contagion models per se.

They emphasize that the model predicts as much as 30 out of the 1,000 people in our population will die in a matter of a few months. That is ten times the average historical rate for all causes of death by disease. So everyone takes notice of the dire prognostication.

Now suppose that the elders, in a stark departure from long-standing libertarian traditions, suddenly command everyone to shelter in their huts save only for necessary outings, such as for gathering staples, seeking medical treatments. But all such outings are to be approved, and their manner specifically prescribed, by the elders, including the mandatory wearing of protective gear or equipment, which is said to be effective despite lack of evidence to support that view.

If protective gear works, why can't people just go where they please? The answer, say the elders, is that people can't be trusted to consistently wear the gear, and that will make it impossible to contain the spread. They also insist that if both the potential receiver and transmitter of the disease agent wear protection it increases disease prevention effectiveness, a claim not based on any repeatable experiment and otherwise not verifiable. Ostensibly for their own protection, people must wear such safeguards the few times they are allowed to be out.

Because fear grips the island as a result of the elders' pronouncements, most people agree to such authoritarian measures, while the rest, the skeptics, carefully try to weigh the costs of heeding versus ignoring those rules.

They find that the quantifiable price of falsely believing that the elders' central claim is true is at least the sum of these: human rights violations, financial impoverishment, physical and mental health degradations from house arrest and social isolation.

But they conclude that there's an unknown cost to falsely believing the elders' claim is incorrect[32], citing the sheer uncertainty of the models and the lack of practical falsifiability of the theory. The cost is capped at the elders' prediction of 300 dead, but only if taken at face value—which is another way of saying the cost is unknown.

Under such uncertainty, rational choice is impossible. No decision as to how to proceed can be based on a theory whose models give wildly different possibilities[33]. So the

---

[32] I.e., the case where the pandemic is real but people think it's not.

[33] For a real-life example, see [42], [43] and [44].

skeptics abandon the scientific question altogether. Instead they frame the problem as a game, where the object is to find a scenario under which people can act according to their preferences without detrimentally impacting others.

They find that the right scenario is just what naturally happens without intervention:

Those who fear contagion avoid human contact, else wear protective gear they believe works[34], in situations where contact is unavoidable. Those who don't fear human contact or don't care about consequences to themselves just carry on as they normally would. Problem solved without violating human rights, and without undue economic, physical and psychological suffering.

Considering that illustrative perspective, which intentionally parallels the supposed virus allegedly ravaging the world in 2020, it should be clear that there was never any logical justification for violating the human rights of individuals based on that pandemic narrative. And we concluded this by simple logic and common sense, without need of science at all.

---

[34] In this analogy intended to mirror real life events, the elders' belief that protection is necessarily two-way is based on faith or mere assertion, rather than reason.

But it would be useful to know if the pandemic notion has scientific merit. For instance, if the scare turned out to be false it would simplify everyone's life and end the suffering exacted by government measures. If however it turned out to be true, it would compel those who otherwise were skeptics to take measures to protect themselves if they cared about their own lives.

So what I will show next is that, with a little more effort, we can take apart the narrative by its own pronouncements and examine the validity of the more technical and scientific claims without being experts in their underlying domains.

JOURNEY THROUGH A PANDEMIC

One day life is normal; the next it's not. What changed? You read in the press that cases of a mysterious new virus are alarmingly on the rise. All major news outlets around the globe are in agreement: we are doomed if we don't take drastic measures to "stop the spread of the virus", because allegedly scientific models of viral spread and virulence predict death rates of epic proportions, to the tune of millions in the USA alone. [45-48]

But neither the number of deaths officially and widely ascribed to the virus nor the rate of increase of official cases suggests anything gravely out of the ordinary for any health condition. And neither are people dying or getting sick in historical disproportion.

The entire pandemic narrative hinges on at least the following claims:

1. A new disease exists.
2. A novel virus causes the alleged disease.
3. The virus is sufficiently contagious that it justifies alarm and invocation of human rights-restricting emergency powers.
4. The widely circulating viral spread model results claiming high infection rates and projecting a huge number of deaths are accurate.

Meanwhile, the popular press gives the impression that health chaos ensues all over the world. Hospitals are over-run with apparent spikes in deaths in key places beyond the ostensible epicenter in Wuhan, China, such as New York City and Lombardy, Italy. [49-52]

It's hard to make any firm conclusions regarding those claims. Whom to trust, citizen journalists with cell phone cameras around the world and independent media reporting that hospitals are apparently *not* at all overrun[35] or the mainstream press depicting quite the opposite? Even the mainstream media appears to be confused, warning that hospitals variously are or will be over capacity while admitting that makeshift supplemental field hospitals have gone unused [53-56].

Scientific and other technical claims can be another matter entirely. Eyebrow raising observations, for example, surfaced which strongly discredit the narrative, such as the CDC[36] and the WHO[37] separately publishing guidelines effecting an unscientifically low standard for concluding death-by-COVID-19––literally directing healthcare

---

[35] Many private individuals and small independent media sources independently took to filming medical facilities and interviewing hospital staff during the same time that mainstream outlets claimed that those very hospitals were being overrun with pandemic patients. These videos suggest that the hospitals were in fact quiescent. As of this writing several associated videos were taken down by YouTube (such as [57] and [58], which remain mirrored on a different platform as of this writing) and a campaign to discredit alternative views on the matter has been well under way [59], but many of those accounts may still be found under the #FilmYourHospital Twitter hashtag.

[36] "Covid 19 should be reported on the death certificate for all decedents where the disease caused **or is assumed to have caused or contributed to death.**" (bold emphasis theirs) [60]

[37]"Both categories, U07.1 (COVID19, virus identified) and U07.2 (COVID19, virus not identified) are suitable for cause of death coding." See especially item "3" on page 3 in [61].

providers to mark death certificates as being caused by the alleged pandemic virus from mere suspicion of infection, irrespective of any preexisting disease or condition. It is not an exaggeration to say that those guidelines render meaningless the number of deaths officially ascribed to the putative virus.

And in case you didn't read those guidelines on the CDC and the WHO websites, Anthony Fauci[38] and Deborah Birx[39], respectively, could be seen and heard advocating complete concurrence with that policy. They did so on live national television, offering the unabashed view that dying *with* the virus must naturally be treated as dying *of* the virus, no matter what underlying condition the patient had.

That's like saying a guy who got shot through the head *with* a cup of coffee in his hand died *of* the cup of coffee. It's like marking the cause of death as "the virus" for a guy who came in with an underlying heart condition who then died of a heart attack after coincidently testing positive for

---

[38] "I can't imagine if someone comes in with coronavirus, goes to an ICU and they have an underlying heart condition and they die [that] they're going to say 'cause of death: heart attack" [62].

[39] "If someone dies with covid 19 we are counting that as a covid 19 death." [63]

said virus[40]. It makes a mockery of something as solemn as determining and recording cause of death. Indeed it makes a mockery of the word science, and the faculty of pathology is ostensibly founded on it. But those have been literally the official guidelines worldwide.

It's enough to reflexively shake your head and ask, is this for real? Isn't this fraud? If the deaths data can't be trusted, why is the mainstream press acting like it's all kosher? Why would the CDC publish guidelines patently advocating pseudoscience and fraud and then follow up with illustrative examples of same in televised forums beside none other than the President of the United States, as if to gaslight[41]?

At the same time the WHO's guidance to healthcare providers states, "Virus isolation is not recommended as a routine diagnostic procedure," [65]. As we'll see later, no researchers have isolated an actual virus associated with the

---

[40] That's actually a paraphrase of Dr. Fauci's own example. [62]

[41] Some posit that this is deliberately induced cognitive dissonance as an attempt to destroy the listener's ability to trust his own intellect, leading him to seek solace in the official word. See the works and lectures of Soviet defector Yuri Bezmenov, sometimes found under the pseudonym "Thomas Schuman" [64], for related KGB tactics. I do not claim U.S. officials deliberately subvert by gaslighting, only that it's a possibility worth pondering.

alleged pandemic disease. So it's rather odd that the WHO would discourage proper investigations that could support its contention of the existence of a novel contagious disease of pandemic proportions.

Nevertheless, authorities push the WHO's narrative that a novel disease-causing virus exists. Why would authorities insist that a new, pandemic-level, highly infectious viral disease exists while at the same time discourage proper investigation of it? Without such rigor, how could medical professionals feel sure that the apparent fatality spikes in places like New York City and Lombardy, Italy, were not caused by something other than a novel virus?

There are no ready answers to these questions, and the press doesn't address them or doesn't make an issue out of them. Yet there it is in plain sight, authorities themselves openly directing hospitals to improperly ascribe cause of death to the alleged virus and then turning around and openly using the resulting meaningless tally as justification for tyrannical rule in the guise of pandemic countermeasures.

Faced with this absurdity, there is rational justification for suspecting that the four scientific claims above may be utterly false. In fact, it's fair to say that we would be hope-lessly naïve if we didn't at least suspect as much.

GOING TO THE SOURCE

If a novel virus has caused novel disease researchers must have isolated it, taken its picture, examined its genetic code, kept it in a jar, so to speak, or otherwise showed di-rect evidence that it exists and that is has communicated specific symptoms from one person to another. It's only common sense.

There are key peer-review papers in circulation claiming that the virus was in fact isolated and that it caused the al-leged disease. These are the very papers upon which the entire premise of the pandemic originally hinged[42]. And given what's a stake, it's worthwhile to read them.

But before doing so, the first questions to ponder are at least these: How do we know that the suspected pathogen caused disease? What specialized scientific methodology is logically required for that with viruses? Are we, as non-practitioners of molecular biology, microbiology or other relevant sciences, able to judge whether the procedures performed warrant the peer-reviewed authors' conclusions

---

[42] [66-76]

and wide acceptance by healthcare institutions and political powers?

There is indeed appropriate method in this case, and it's fully accessible to the lay reader. We are free to scrutinize it as we would anything else. It exists in the form of a general procedure known as Koch's postulates [77]. I'll revisit them in a moment, but first let's view the viral pandemic claim from the perspective of pure common sense.

We want to know if a hypothesized agent (an alleged virus in this case) solely[43] caused a set of symptoms of specific disease. We further want to determine, in that event, if the condition demonstrated communicability by at least the normal course of human interaction.

We need not be vaunted virologists to understand this simple causality proposition. The idea is that the alleged virus jumped from one person into another and caused the

---

[43] Any claim that a pathogen *alone* causes disease summarily discounts other potential or synergistic causes, such as the state of the host and environmental conditions. It could be that introduction of a pathogen results in disease only in the event that the host's contextual state—nutritional, atmospherical, genetic—is conducive in some way. Koch's postulates are not designed to address this view, and neither is allopathy's Pasteurian philosophy. But I'll leave that issue aside for the sake of brevity, and because it won't impact conclusions made in this chapter.

latter to acquire the former's same illness. That's at the heart of germ theory.

To verify that proposition we would naturally reason along these lines: if we could somehow identify and separate—a.k.a., isolate—the agent from all other potentially confounding substances[44] and follow its path (note its presence) we could see if it did in fact jump from one person to the next and if its arrival consistently preceded the onset of observed symptoms in the absence of other causative factors.

But there's a problem. Viral theory holds that a given virus affects people differently. Some people get sick from it; some don't. Consequently the biomedical establishment's view on coronavirus is that people can be asymptomatic carriers. Looking and feeling healthy doesn't imply not being lethally contagious. Everyone is potentially an unwitting seed of death. There is no way to predict who is and who isn't, because a person not infected today may be a carrier tomorrow, without the slightest indication.

---

[44] Any other substances either known to cause disease or not sufficiently understood to rule them out. We wouldn't consider pure H_20 to be a confounding substance, for instance, but we would a bacterium previously proved to have caused disease, or any substance known to be toxic.

The theory offers no cure for this ambiguity, but it has served as justification for quarantining the healthy—tantamount to what would ordinarily be considered illegal, arbitrary detention (house arrest). It's also curtailed the act of being human; people can't see each other's facial expressions in public; fear keeps friends and family members apart; children can't play together in school and are robbed of the warmth of human interaction.

The concept of asymptomatic carrier is clearly very profound; it renders everyone a potential threat and results in the abrogation of civil liberties and even basic humanity. We therefore should not take it at face value if we care about such things. So let's first see if it's falsifiable:

Supposing coronavirus theory were false, how could we prove it so if the physical presence of the posited pathogen doesn't even imply illness? We clearly can't. The asymptomatic carrier paradigm precludes any experiment showing that coronavirus does not cause illness.

Therefore, coronavirus theory is not strictly falsifiable[45]. More generally, any virus professed to spread

---

[45] As I will discuss in a moment, in principle one could formulate a coronavirus theory that would render it *statistically* falsifiable. But as we'll see later, this isn't the case in practice.

at least partially by asymptomatic contagion would similarly be an unfalsifiable theory. Does that mean such a theory couldn't possibly be true? No. Falsifiability doesn't test that. It just checks if something admits a test of falsity. But without falsifiability we would have to take the theory on faith.

For such a theory to be falsifiable it would have to predict in what particular circumstances the virus doesn't cause disease. Otherwise the theory's proponents could always say the absence of disease in any particular instance was due to something unknown but that the theory is still true.

Not all is lost, however. We can still test the theory in some useful way. If we could count how many times disease wasn't conferred by the virus in a random group of test subjects, as compared to a control[46] group, we could gain various degrees of confidence of the likelihood that the virus isn't the cause of disease. In that sense, we would have a statistical falsifiability principle.

For example, say we theorized that a particular particle P causes P-disease. Suppose that out of a thousand ran-

---

[46] A control group in this context is a randomly selected number of subjects not exposed to the virus. See *Appendix ii*.

domly selected people (test group) we find on average that 17 develop the same illness when exposed to *P*, whereas only 3 in a random control group of the same size become ill.

Further suppose we can calculate that the probability of getting that difference by pure chance is 0.1%[47] (or equivalently, 99.9% confidence that the observed difference was not by chance).

We could then define *P*-disease theory as one which satisfies two criteria: A) statistical significance (greater than 99.9% confidence) that exposure to *P* results in B) at least 14 more ill people per thousand than in its absence.

That way it would be possible to prove the theory statistically false, for example, if experimentation showed either A) less than or equal to 99.9% confidence of a non-chance result or B) that the difference in the number of people falling ill between the test and control groups was less than 14 out of 1,000, under any level of confidence.

---

[47] This is calculated, for example, by means of the *t*-distribution, which comes from counting how many ways we can expect, by pure chance, to have a given difference of (e.g., 14) or larger between two group averages. [151]

Is this the general approach taken by the medical establishment with regard to coronavirus theory? Before answering that, we need to consider Koch's postulates [78].

KOCH'S POSTULATES

1. The microorganism[48] must be found in abundance
    in all organisms suffering from the disease, but
    should not be found in healthy organisms.

2. The microorganism must be isolated from a dis-
    eased organism and grown in pure culture[49].

3. The cultured microorganism should cause disease
    when introduced into a healthy organism.

4. The microorganism must be re-isolated from the
    inoculated, diseased experimental host and identi-

---

[48] Note that Koch's postulates predate viral theory, and the biomedical establishment does not consider viruses to be microorganisms. This has led some to conclude that the postulates don't apply
to viruses. The problem with that view is that "microorganism" is
an arbitrary distinction. The study of causality, which the postulates
embody and our common sense can parse, is independent of the
taxonomy assigned to the entity under study.

[49] In the case of bacteria, this is unambiguous. But for viruses it
requires the belief that they reproduce by infecting bacteria, so that
culturing bacteria also cultures viruses. Be aware though that this
parasitic relationship is a theory which we would ideally need to
test. But I will proceed by assuming it's true, because it won't affect
conclusions.

fied as being identical to the original specific causative agent.

Note that what the postulates propound is purely common sense:

If we suspect that a virus is the sole cause of a particular disease, at the very least we ought to physically take it from a diseased host and then introduce it to a healthy one to see if the illness is replicated, accounting for any other known disease-causing agents. Doing so would imply first identifying and isolating the virus (separate it from everything else that could confound results) so we knew it really was there and that we really were transferring it to another host. What remains is then to see if it causes disease—i.e., the third postulate.

Because viral theory, as you recall, posits that sometimes a virus inexplicably does not result in disease, any claim of viral disease is relegated to being at most *statistically* falsifiable. This means that we can only gain varying degrees of confidence that a suspected virus either caused a particular instance of illness in the past or that it never causes that illness[50]; the more instances the more confidence in either.

To shed more light on the reasonableness of Koch's postulates, what follows is a hypothetical example of one such instance[51].

A medical researcher hypothesizes that a disease is communicable and is caused by a certain agent—a particle having certain identifiable characteristics. So he takes fluid samples (say from the lungs or nasopharynx, if the symptoms involve the respiratory system) from a suspected host exhibiting a set of very specific symptoms associated with the disease under study.

From those samples he isolates (Postulate 2) the particle by completely, carefully and meticulously filtering out all other material in the fluid [79]. If that particle is common to healthy hosts (Postulate 1), he stops and concludes that

---

[50] Recall that falsifiability comes from the idea that we can only prove things about the past, not the future. So a statistical context can only change our confidence that a result *was* a true prediction versus mere coincidence.

[51] Modern medical science considers the observation of a single instance to be sufficient. However, per the discussion in this chapter, it should be clear that a single experiment can't by itself provide any statistical measure of confidence in the conclusion drawn.

it could not cause disease; otherwise he takes the particle and introduces it into a new host.

If the new host exhibits the same symptoms as the original host (Postulate 3), and if the researcher can isolate the particle from the new host anew (Postulate 4), to verify it was present during the expression of those symptoms, the researcher concludes that the agent could *possibly*[52] have caused the disease. (If he subsequently gets this same result more than would be expected by chance[53] after repeating the experiment more times with new hosts, he gains more confidence that the agent was causal to disease in those instances.)

Finally, successful reproduction of the symptoms in the above way doesn't logically establish contagion per se unless the experimenter introduces the agent from host to host in a way that reflects the same hypothesized communicability of the alleged disease.

For example, injecting the agent with a syringe is not something that happens during, say, a conversation be-

---

[52] I say "possibly" because correlation does not imply causation. However, repeated experimentation where other factors are accounted for increases the confidence that the experiment was an observation of a causal event.

[53] Vis-à-vis the control group.

tween two people. So if he's looking for contagion resulting from normal human interaction, he wouldn't involve injection as part of the disease transfer protocol of the experiment. This is merely common sense.

SEMINAL PAPERS

We are now well positioned to scrutinize relevant coronavirus foundational peer-reviewed articles claiming to have established a causal link between the alleged virus and its presumed disease. These are the only relevant scientific publications available as of this writing and most of these were in press at the time of the health emergency declarations by governments around the world in the name of the putative pandemic virus of 2020. Therefore, the bulk of these papers unequivocally forms the ipso facto basis for government positions and actions in ostensible response to the virus.

I will however spare you the page-by-page technical scrutiny of each one of those studies, as that would fill up its own (pedantic) book-scale volume. I will instead summarize the findings below and then parse one representative publication in the next section.

Ideally, if you can spare the time, read the papers yourself, keeping this discussion in mind before going further. Depending on your background, you may need to reference some biological or biomedical research terminology, but there's otherwise nothing insurmountable in that literature as far as comprehending the important points.

A list of all the relevant peer-reviewed articles published around the time of the alleged pandemic can be found in the bibliography. [66-76][87]

After reading them, I highly recommend that you follow the excellent analysis by Dr. Andrew Kaufman on this topic entailing several of these very papers, if it's still up on his YouTube channel or on his website while this book is in print[54]. Despite being directly qualified to address the topic—his background is in molecular biology (Massachusetts Institute of Technology) prior to his M.D. degree—he speaks plainly and conveys the relevant points without needless jargon. [81]

---

[54] If not, search directly for "The Rooster in the River of Rats", the title of his video, on the web or other video hosting websites, such as Bitchute. Failing that, other similarly qualified individuals have conducted similar analyses; see for example [82].

Before going further keep in mind that, because of unknown unknowns[55], observation of an association between virus and disease may imply causality only if every other possible causes can be ruled out. But it's not possible to rule out what is unknown. That is the reason for adopting a statistical paradigm[56], with control groups and confidence levels, as previously described.

Similarly, the asymptomatic carrier assumption predicts unexplained negative postulate results for at least some hosts. This would imply unknown unknowns accounting for mechanisms that render some people impervious. Consequently the only conclusions available in this context are also statistically qualified ones, requiring the usual test and control groups, respectively, with as large a population size as feasible.

But not only did the authors not perform the postulates in the above statistical sense, they performed literally *none* of them in any sense, even though some authors claimed to have actually fulfilled part or all of them. Despite some study authors' own claims to the contrary, not one paper even attempted to isolate a virus nor measure

---

[55] See alpha world and related discussions in Chapter I.

[56] I.e., *pcrdb* (see *Appendix ii*).

disease in situ, two common sense themes central to Koch's method.

These are serious shortcomings that even non-experts in the field can attest. If a virus was never isolated, how can they claim, as they have, to know what RNA sequence it would have? If the RNA sequence isn't actually known, how can anyone construct, as the U.S. CDC and the WHO have, respectively, an RNA-based test for such a virus?

It turns out that the RNA composition was originally concluded based on examination of patients from Wuhan, China, the declared epicenter of the alleged virus, where investigators took lung fluid samples and performed what is known as Random RT-PCR[57] to deduce the presence of

---

[57] Random RT-PCR (Reverse Transcription PCR: https://en.wikipedia.org/wiki/Reverse_transcription_poly-merase_chain_reaction) is the process of testing for the presence of random RNA base sequences in a biological fluid sample. If the sample contains strands having only one unique sequence, one can in principle deduce it, at least in part, by separately testing for a large enough number of smaller sequences. For example, if the unknown strand is GAUUACA, a test for GAA would be negative and tests for UAC and ACA would be positive. Because the latter two positive tests overlap to form UACA, we conclude that UACA is part of the unknown sequence (like trying to deduce someone's name by stitching and overlapping together random, correctly guessed character sequences). Again, this approach assumes the biological sample contains only one unique sequence. Thus the greater the number of unique sequences the more ambiguous the results.

an allegedly novel contiguous sequence of RNA bases forming ostensibly only part of the proposed viral genome. From that they deduced that this RNA sequence must be from a virus, and a novel one at that. [68]

The central problem with that conclusion is that body fluid routinely contains RNA from many sources: host cells themselves, bacteria, fungi. As I will further expound in the next section, there is no basis for believing such RNA is from a novel virus, let alone one which causes disease. Even if that RNA were from a virus, a PCR[58] test can only measure RNA, not viruses. The presence of RNA matching part of a hypothesized virus is not the same as the presence of an actual virus.

Nevertheless, the CDC's and the WHO's RNA based tests have been widely applied as the gold standard for determining who is infected with the presumed virus. But obviously, if the RNA is of unknown origin, those tests are utterly meaningless.

TOO LITTLE, TOO LATE

It's interesting to note that it wasn't until June 6, 2020, months after the presumed novel virus allegedly ravaged

---

[58] [83][84][85].

the world, that the CDC deigned to publish a research paper claiming to have isolated and sequenced such a virus[59]. As previously mentioned, and as will become more evident in the ensuing passages, isolation is absolutely essential to experimental investigation as to causality.

But in short, the CDC paper did not isolate any such virus. The key passage is the *Cell Culture, Limiting Dilution, and Virus Isolation* portion of the *Methods* section.

I will summarize the authors' procedure in three points (NOTE: content inside parentheses in this list below denote specific nomenclature used in the paper, which I later highlight in bold in a quote of a relevant passage):

1. A collection of living human and non-human animal cells (known as a *culture*) are pretreated with antibiotics and anti-fungal medication and then inoculated[60] with fluid samples ("swab specimens") from a patient suspected of containing a disease-causing virus.

---

[59] [86][87].

[60] In this case "inoculated" means that the authors introduced allegedly diseased human bodily fluid into the collection of living animal cells.

2. The animal cells, which are in a nutrient medium so they don't perish due to lack of sustenance, are then observed (via "standard plaque assays") for the appearance of any cell deaths ("cytopathic effects").

3. If cell death is detected, an enzyme ("viral lysate") is introduced to the cell culture to break cell walls, releasing genetic material (nucleic acids). The RNA in that genetic material is then sequenced and *assumed* to be part of a viral genome.

In other words, they took swab samples from a patient and introduced them into medicated animal cells to see if any of those cells died. Because they observed that some cells did die they concluded—falsely, as I will show—that a new virus exists and caused those cells to die.

They then searched for unique RNA sequences in the inoculated, medicated cell culture media, by using enzymes which break down cell walls, releasing nucleic acids (genetic material, including RNA). They assumed—again falsely—that those sequences belong to a novel virus responsible for the observed cell deaths.

Here is the relevant passage in the *Methods* section (NOTE: don't be put off by unfamiliar technical terms; "Vero E6, Vero CCL-81, HUH 7.0,..." below, for example, just refers to an assortment of human and non-human animal cell lines making up a cell culture):

We used Vero CCL-81 cells for isolation and initial passage. **We cultured Vero E6, Vero CCL-81, HUH 7.0, 293T, A549, and EFKB3 cells in Dulbecco minimal essential medium (DMEM) supplemented with heat-inactivated fetal bovine serum (5% or 10%) and antibiotics/antimycotics** (GIBCO, https://www.thermofisher.com). We used both NP and OP **swab specimens** for virus isolation. For isolation, limiting dilution, and passage 1 of the virus, we pipetted 50 μL of serum-free DMEM into columns 2–12 of a 96-well tissue culture plate, then pipetted 100 μL of clinical specimens into column 1 and serially diluted 2-fold across the plate. We then trypsinized and resuspended Vero cells in DMEM containing 10% fetal bovine serum, 2× penicillin/streptomycin, 2× antibiotics/antimycotics, and 2× amphotericin B at a concentration of 2.5×105 cells/mL. We added 100 μL of cell suspension directly to the clinical specimen dilutions and mixed gently by pipetting.

We then **grew the inoculated cultures** in a hu-
midified 37°C incubator in an atmosphere of 5%
CO_2 and **observed for cytopathic effects** (CPEs)
daily. We used **standard plaque assays** for SARS-
CoV-2, which were based on SARS-CoV and Middle East respiratory syndrome coronavirus (MERS-
CoV) protocols (9,10).

When CPEs were observed, we scraped cell mono-
layers with the back of a pipette tip. We used 50 µL
of **viral lysate for total nucleic acid extraction
for confirmatory testing and sequencing.** We
also used 50 µL of virus lysate to inoculate a well
of a 90% confluent 24-well plate.

As you can see from that methods section, the authors did
not physically isolate, or purify, a virus from other, potentially confounding substances in solution, nor did they attempt to visually identify one, such as with an electron mi-
croscope.

How could they claim to have sequenced the RNA of
such an entity? How could they know the origin of such
RNA? That would be like concluding that all the ticket

stubs found on the floor of a concert hall originated from purses, without seeing any purses.

It's not logical to conclude from the experiment that RNA found in the medium is necessarily of viral origin because, as previously mentioned, there are many (non-viral) potential sources of RNA in such a medium.

For example, cells may release vesicles known as *exosomes* in response to stress from toxic substances such as the very antimicrobials used in the experiment[61]. By definition, viruses and exosomes are structurally identical: RNA wrapped in a kind of cell wall material. Therefore, how did the experimenters conclude that applying *viral lysate* to dissolve such walls did not result in exosomal RNA release?

But even in the absence of that ambiguity, there could still be other sources of non-viral RNA. Normal host cell turnover arguably results in RNA release; the bacteria-killing antibiotics used in the experiment logically result in the release of bacterial RNA; and similarly for antimycotics vis-à-vis fungi.[62] The authors take no account of those sources. Some bacteria and fungi may already be dead in the fluid, which would result in RNA release.

---

[61] See for example [90].

[62] None of these examples is scientifically controversial.

One also can't logically conclude from the experiment that a virus was necessarily responsible for the observed cell toxicities. The authors' treatment of the cell lines with antimicrobials could have been at least one cause of cell deaths[63].

These confounding experimental considerations and potential artifacts would have been avoided if the authors had followed the spirit of Koch's line of reasoning. Instead the end result is that the experiment is entirely inconclusive.

It's worth mentioning that researchers have offered alternative motivations for concluding the existence of a novel, disease-causing coronavirus. One is the observation of relatively high RNA similarity of the allegedly novel virus to that of a supposedly preexisting virus known as SARS-CoV-1, suggesting a new pathogen in the coronavirus family, such as described in Zhou et al [69].

But the problem with that reasoning is at least twofold: 1) the conclusion that the original SARS-CoV-1 is a disease-causing pathogen itself suffers from the very same experimental shortcomings I describe above [72]; and 2) the Zhou paper claims a genetic similarity of 79.5% while

---

[63] [88][89].

humans and chimpanzees ostensibly have a genetic similarity of 96% [153], raising the question of meaningfulness.

What is the appropriate similarity metric? What degree of similarity differentiates "virusness" from "exosomeness" or one viral "family" from another? Would having an RNA segment less than 79.5% similar to an allegedly known virus mean that the RNA is of something non-viral? What would be the basis for that cutoff? What determines that a given RNA sequence is viral in the first place? It would therefore seem that the Zhou et al paper is inconclusive at best.

But what about the fact that several papers alleging a novel coronavirus report the same RNA sequence associated with separate patients exhibiting the same symptoms? Would that not indicate a common viral cause?

As we saw earlier in the analysis of the CDC paper, RNA has many sources. One source may be human cells themselves. Finding the same sequence in different people is possible without invoking the existence of a virus. The sequence could simply be common to people's genomes, for example, or it could come from common flora (bacteria, fungi).

With regard to the scientific method, the above issues beg the question of, why would experimenters forgo the more direct investigative route offered in Koch's postulates or similar logic in favor of proximal arguments—such as the above—fraught with ambiguities and approaches that avoid more direct study of causality?[64]

And yet this is the choice authors made in the case of the alleged, novel coronavirus. That is like concluding that raccoons have been stealing your chickens because raccoons possibly dwell in the area, rather than setting up observation points to catch them in the act. Just because your chickens were taken does not imply the raccoons did it, as plausible as that may be. Plausibility does not amount to a scientific conclusion.

So there it is in full grasp by non-experts in the field of germ theory... There is as yet no scientific support for the belief that a novel disease-causing virus even exists[65]. And we were able to make that determination from sources that included the horse's mouth, as it were: the very medical

---

[64] Is it because researchers haven't been able to fulfill Koch's original postulates for viruses generally?

[65] See also the 13 July 2020 CDC admission that "...no quantified virus isolates of the 2019-nCoV are currently available." [https://www.fda.gov/media/134922/download]

establishment credo as published by the US Government's own ministry of health, the Centers for Disease Control and Prevention.

Without scientific support for an actual virus, the notion of a novel infectious disease is not credible.

Without scientific support for the existence of a virus there can be no credible clinical diagnostic test ostensibly detecting it, no matter how many physicians and virologists may suggest to the contrary in the mainstream press or the local healthcare facility.

Without a clinical diagnostic test, discussions on infectious disease spread characteristics, case fatality rates, face mask effectiveness, infectiousness of the alleged disease, concoction of vaccines, etc., are completely without meaning.

It is not possible to estimate the threat posed to society by an unfounded disease. It would be pure make believe. No dire warnings of people imminently dropping like flies if they don't comply with "pandemic countermeasures" have any credibility in this context; and yet that did not stop authorities from prognosticating health pandemonium.

Without a believable threat of such magnitude, governments have no basis for declaring health emergencies and no pretense[66] on which to predicate the onerous and draconian measures people have had to endure to date.

If the alleged infectious disease actually existed, why would the WHO and the CDC feel the need to take deliberate steps to, in effect, induce fabrication of case numbers by publishing guidelines[67] requiring healthcare providers to mislabel death certificates with the alleged novel disease as the cause of death?

A CYNIC'S SUMMARY

A virus never isolated (and thus not shown to exist nor cause disease) but somehow infecting people, the majority of whom suffer no symptoms, presumably pass on the speculative virus to others, the vast majority of whom also suffer little to no symptoms ascribed to said virus[68]. This yet-undetected virus is said to cause illness predominantly or almost exclusively in those who were already sick and

[66] See previous discussion in Anatomy of a Pandemic concluding that, even in the presence of a pandemic there would be no logical justification for dissolving civil liberties.

[67] As previously referenced, see [60][61].

[68] The WHO estimated the number at 80% [91].

dying from one or more prior serious illnesses and old age[69].

Because all such deaths are nevertheless included in the overall viral death tally, physicians are to tacitly accept that demonstrably dying of an underlying condition in no way exculpates the imagined virus. Therefore authorities require the attending physician to mark the death as being caused by this virus if the physician at least suspects it somehow played a role in the patient's death. No need to investigate cause of death by trying to isolate a putative virus from the patient or ruling out other causes. The physician's hunch will do just fine, especially if his healthcare employer is financially incentivized[70] to see a causal relationship.

The presumed viral death toll in turn motivates and justifies the implementation of government countermeasures such as the following:

If you don't wear a mask not shown to protect against viruses or don't socially isolate on account of a virus not seen nor characterized which kills ostensibly only those al-

---

[69] See for example [154(a-b)] and [155].

[70] As of this writing, the U.S. Centers for Medicare & Medicaid Services (CMS) pays out $13,297 per patient diagnosed with the speculative virus and $40,218 if the patient is placed on a ventilator for more than 96 hours. [92]

ready dying of other causes, you deserve a fine, imprisonment or both for (not) putting at risk those already on
death's door.

Even if you wear a mask, you don't deserve to operate
your small business if it caters to the general public, except
when the government arbitrarily deems that the infection
spread curve looks "good enough", as assessed by means
of a demonstrably meaningless test. However, you can't
allow more than typically 25-50% customer capacity, depending on your jurisdiction; you must also be ready to
shut it down at a moment's notice by government decree.

If you have any questions about the science, the answer
is... Trust the evidently ethically-challenged[71] official agen-
cies and their affiliated or media-christened expert organizations, such as the US CDC and part privately-funded[72]
World Health Organization.

And you should trust your physician, unless of course
s/he disagrees with the establishment narrative with regard

---

[71] [145][146].

[72] The WHO itself reports that the Bill & Melinda Gates Founda-
tion was the second largest donor to the WHO in the 2018-2019
cycle, behind the United States and just ahead of the United Kingdom of Great Britain and Northern Ireland combined—p. 82 in
[144].

to the alleged virus; in that case s/he deserves to be brand-
ed an anti-vaxxer[73], because the intent of Big Pharma is to
vaccinate everyone based, incidentally, on untested novel
theories[74] only now being evaluated in an accelerated[75]
fashion, no less, forgoing proper scientific inquiry in the
name of an unfounded health emergency, with the promise
that life will go back to normal[76] without going back to
normal[77].

---

[73] The tenor of the mainstream press has the effect of discrediting
anyone questioning the push for the anticipated coronavirus vaccine and vaccines in general. A typical example, from CNN, puts all
critics in the same category as those claiming that the vaccine is a
"CIA plot to take over the world," which is a transparent evasion
around reasoned discourse based on the well known *straw-man* tac-
tic. [156]

[74] So-called RNA and DNA vaccines, which allegedly instruct the
host to inoculate itself by producing antigens in its own cells, involving the introduction of foreign genetic material into the host's
cell nuclei. See for example [93].

[75] See Coronavirus Treatment Acceleration Program of the U.S.
Food and Drug Administration (FDA) [94].

[76] [147][148].

[77] [149][150].

III. Herd Mentality

COMMON GOOD V. COMMON SENSE

Herd immunity is the idea that those for whom vaccination does not confer immunity are de facto protected by those for whom vaccination does [95-98]. If true, it implies that vaccination is not only for one's own good but also for the *common* good.

In 1905 the Supreme Court in *Jacobson v. Massachusetts* [99] decided that a person may undergo forced vaccination for the common good, citing the herd immunity principle.

The case decision was ultimately only to require that Jacobson, the plaintiff, pay a five-dollar fine for refusing to undergo vaccination for smallpox; a hefty sum at the time, but a far cry from the gravity of forced vaccination. Nevertheless, the case would later be successfully cited in *Buck v. Bell* as justification for forcible sterilization of an individual [100].

Considering those as-yet unchallenged court decisions in the backdrop of a long U.S. tradition of enacting compulsory measures such as military drafts, jury service and childhood vaccination as a requirement for school attendance, it does not take a legal scholar to see the potential for compulsory vaccination during proclaimed health emergencies.

But here is one renowned Constitutional scholar and acclaimed trial lawyer, Harvard Law School emeritus professor Alan Dershowitz, who publicly proclaimed that *Jacobson v. Massachusetts* serves as full legal precedent supporting the doctrine of forced vaccination during a pandemic [101]. As did the judges in Jacobson v. Mass, Dershowitz implicitly assumes the infallibility of the medical establishment. To Dershowitz and the 1905 Supreme Court, the notion that herd immunity, let alone vaccination, could be false or unscientific is evidently not worthy of consideration.

Herd immunity plays a prominent role in the vaccination narrative most of us have taken for granted since adolescence. It's not an exaggeration to say that it is a sacred cow. The principle is as well regarded and as firmly planted in the modern world as is vaccination itself, thanks to the advent of allopathic medicine out of Europe and the United States in the early nineteenth century [102].

In establishment medicine and popular culture, dissent against such vaunted principles is equivalent to asserting that the sky isn't blue or that the Earth is flat. It's patently absurd to publicly suggest that herd immunity is false, let alone that vaccines don't work or are unsafe. It would be an act of pure heresy.

Betting against a world-wide establishment credo of this magnitude is gravely ill-advised. If your staunch intent is to invite ridicule or worse, I would strongly encourage you to publish anything questioning it.

If we adopt the establishment posture we would say that obviously vaccines save lives and that obviously there wouldn't be an apparently dominant number of scientists and healthcare professionals adhering to the vaccination ethos if it weren't true.

But with that in mind, consider the following points.

The once common medical practice of intentional, induced bleeding likely killed George Washington in 1799

[103], at a time when the medical establishment was certain that bloodletting cures illness [104].

Abraham Lincoln is believed to have suffered terribly from a blue pill elixir containing mercury as the active ingredient, in his vain attempts to control migraines [104], but medical orthodoxy from the 18th to the early 20th century had it that mercury based medicines were good for treating headaches and other ailments.

Hundreds of thousands of men, women and children lost their lives from the mayhem and chaos following the invasion of Iraq early in the first decade of the twenty-first century by U.S. led coalition forces. But thanks to assurances from Donald Rumsfeld and other White House staff, Saddam Hussein undoubtedly possessed weapons of mass destruction and was intimately connected to the infamous events of 9/11, justifying military invasion of Iraq. [105]

Everyone who was anyone between the second and sixteenth centuries (1400 years!) *knew* that the world revolves around the Earth according to Ptolemy's geocentric model, the notion that the Earth is stationary and that every other celestial body in the universe revolves in literal circles around it[78].

---

[78] [106][107].

In light of the historical evidence, does it follow that there is never any reason to question the establishment narrative? Does it follow that the word of the establishment couldn't possibly lead us astray in profoundly impacting and sometimes lethal ways? Clearly it doesn't. So let's continue with our critique.

CURING THE HEALTHY

Herd immunity, which was central to the Supreme Court's aforementioned common good argument, is said to occur if the vaccination rate surpasses some threshold, ostensibly 70-90%, depending on the contagiousness[79] of the presupposed virus targeted by the vaccine. That appears to be a falsifiable hypothesis. If we observe an outbreak in a population with a vaccination rate at or above that threshold, it would prove herd immunity false, at least for the particular types of vaccines involved in that event.

In fact, there have been many occasions of separate outbreaks in fully vaccinated[80] populations which defied explanation. These involved at least six prominent examples since the nineteen-eighties. [108-113]

---

[79] [95-98].

[80] At least achieving the aforementioned herd immunity threshold.

But that suggests there may be a problem with the vaccination principle itself, not just herd immunity, at least with regard to the specific types of vaccines represented.

Vaccination is the idea that you need fixing before you're broken, so to speak. Every student attending school, at least in developed countries, must accept it as precondition for matriculation, with rare exception. Many parents don't object to it, because it's an ingrained idea since at least the 1930s. But what if it turns out to be harmful, or what if it didn't work? Would we be able to know?

As usual, the key to ultimately answering such questions lies in first determining if the theory admits a test of its own falsity. In other words, is it falsifiable?

But before we can attempt to address that we must be clear about what the theory claims.

Vaccination is said to confer illness prevention, or a kind of protection from breakage. If you were a babushka doll, we could define *craccination* as that which prevents your cracking after being dropped from at most the height of a typical toddler. I could drop you and similarly protected and unprotected dolls on the floor to see which group was more likely to break.

For that theory to be falsifiable we would have to specify just how much protection craccination confers, statistically. The theory would read something like this: Craccination is a procedure which results in the protected being at least three times more likely, on average, to avoid cracking than non-craccinated cohorts when dropped from some specific height or below.

If after we conduct a large number or experiments we don't get the above statistical result, we would logically conclude that the theory is *likely* false. The more experiments confirming that or more extreme result, the more confident[81] we would be of that conclusion. But it would take an infinite number of experiments to strictly prove the theory false, because only that could give us 100% confidence that the average result happened by chance. So in practice we accept that we can only have varying degrees of less-than-100% confidence that we have proved a statistical theory false.

Conversely, if the average result is in line with prediction, we are limited to gaining various degrees of confi-

---

[81] Confidence, or statistical significance, depends on the number of experiments and the separation between the result and what we would expect by chance

dence that the *particular prediction* was not due to chance. We don't derive any measure of confidence as to the trueness of the theory. Recall from Chapter I that in our world we can only prove the past, not the future. This is no different in statistical contexts.[82]

This is the nature of falsifiability for statistical theories. It's not ideal, since proof requires 100% confidence and we would never have that, but it's much better than nothing and is thus useful for our purposes here. Keeping that perspective in mind, let's return to vaccination.

The principle of immunization presupposes that mounting a defense against an invading pathogen requires time for the immune system to recognize the foreign body and time to construct specific countermeasures against it. The theoretical role of a vaccine is to train the body's defense apparatus such that it detects the invader quickly and is ready to fight it off immediately.

An analogy to a standing army might help. We train soldiers that a particular flag or style of uniform signifies

---

[82] In the foregoing example, if you're familiar with basic probability, you'll recognize the explication as equivalent to saying: the bigger the p-value the greater the confidence that the theory as a whole is false (null hypothesis); and the smaller the p-value the greater the confidence that the particular prediction of the theory was not a chance result (alternative hypothesis).

an enemy requiring particular weaponry or strategy to mount a successful defense—call it *battle preparedness theory*––and we give the army that weaponry, in plentiful stacks at the ready. Whereas normally the soldiers would have to wait until the threat is identified, the right strategies devised and construction of the weapons completed, battle preparedness would render defenses immediately ready to defeat the enemy.

Could we ever prove false this general principle of battle preparedness?

Battle preparedness theory generically predicts that the army will win because we gave them the appropriate ways and means to defeat invaders. But there are countless such specific preparations for countless potential invaders, known and unknown. We can't try to disprove every instance without knowing what they all are in advance. Proving false one application of the battle preparedness methodology against a particular invader does not imply that the methodology won't be effective against another. So that implies that the general principle of battle preparedness is not actually falsifiable. Only specific instances are.

If we don't realize that battle preparedness theory is not falsifiable, its proponents could always persuasively explain away prediction failure by saying that the experiment didn't happen to involve the correct ways and means for the particular situation but that the general principle still holds. It's basically an epistemological cheat around observational contradiction.

Unless defined in statistical terms with respect to types of enemies[83], a falsifiable formulation of battle preparedness theory would only be one which specifies absolutely all the ways and means it would confer ability to defeat all invaders, because that's the only form which would strictly admit tests that could prove the entire theory false. It may not be possible to define such a bounded theory in practice, however. But that's the limitation we must live with: If the theory is not properly defined, it is not possible to even *ask* if such a theory is false.

The same is true of the general notion of vaccination. proponents don't define it in statistical terms with regard to the types of diseases it theoretically prevents. The ipso facto definition doesn't claim eradication of, say, at least 90% of targeted diseases on average, which in principle would make it subject to statistical falsifiability.

---

[83] Where battle preparedness would be defined as effective for some statistical fraction of enemy types.

But if proponents of vaccination did propose it in such terms it would require a more complex statistical framework than otherwise[84]. It would also have the effect of creating greater uncertainty, since lumping different vaccine types together results in a kind of averaging effect that loses information[85] specific to each vaccine.

It should be obvious that, when given a choice, it's preferable to adopt the theory formulation whose conclusions have greater certainty. In addition, the larger theory in this case adds experimental complexity to the point where it may render statistical falsifiability experiments impracticable.

So, per that argument, we must treat each vaccine as its own theory. That formulation is (statistically) falsifiable, whereas the other isn't.

To see this, consider that to find vaccination as a whole false requires that all theoretically possible vaccines, not

---

[84] The larger vaccination principle would require testing many populations separately for enough types of vaccines to gain sufficient statistical power. So if a sample of, say, 1,000 test subjects is sufficient to draw reasonable conclusions for one vaccine as its own theory, an overall sample size of 1,000x1,000 = 1,000,000 for a thousand vaccine types may be needed for an experiment testing vaccination as a whole to give similar confidence. This is both more complex and less practicable.

[85] And thus predictability.

just one, fail for all populations receiving them[86]. This is because the falsity of one vaccine for one type of disease——say an alleged flu strain—does not under this formulation logically imply that all other possible vaccines are also false for all applicable diseases.

The problem is that germ theory supposes a countless known and unknown (potentially infinite) total number of pathogen species (the invading armies analogue), each in principle entailing separate vaccines. So the theory is un-bounded. There is no way to test an infinite number of vaccines, nor test ones not yet devised.

Contrast that to the theory of classical gravity. If an object one day rises instead of falls, gravity is false. There is no ambiguity. An alternate theory able to predict which ob-jects rise and which fall would naturally succeed it.

---

[86] Because, by its own definition, vaccination theory relies on herd immunity to protect those for whom it doesn't work, proving a given vaccine false would necessarily require a statistical treatment showing that, despite achievement of at least 70% vaccination cov-erage, at least one member of the population contracted the target-ed disease. This implies a statistical falsifiability treatment whose confidence depends on sample size, degree of vaccination cover-age, fraction who contracted disease, and so on. Refer to Chapter II.

But the emergence of disease in a completely vaccinated population[87] proves false only the implicated vaccine; it does not logically imply that other vaccines can't prevent, or haven't already prevented elsewhere, the same or other diseases. In so doing, the vaccination principle evades falsifiability.

Therefore, the broader idea of vaccination would have to be taken on faith, whereas the idea that a *particular* vaccine works admits a scientific formulation: Vaccine V theoretically prevents disease D; If that prediction fails, vaccine V is false. End of story.

That implies that every time a person contemplates taking a vaccine he or she should logically ask: was that specific vaccine preparation tested by the appropriate scientific method for efficacy[88] involving a placebo control?

It should be self-evident that efficacy would imply testing whether the vaccine prevented disease, but this is not what happens in practice.

The principle of vaccination assumes that artificially stimulating the body's immune system to produce antibod-

---

[87] See the earlier discussion on this topic and previously referenced works [108-113] for many real life examples.

[88] Safety and effectiveness.

ies—called a *humoral response*—targeting antigens confers protection to disease. Consequently, vaccine studies do not test for disease prevention. Instead the credo is that antibody production per se faithfully indicates protection from disease. To vaccination experimenters a successful vaccine is one which stimulates a humoral response, period.

The problem is that the evidence does not support the view that the presence of antibodies implies immunity. Put more rigorously, the theory that humoral response implies immunity is falsifiable and has been unambiguously proved false for at least one disease in mice[89]. That implies that one can't reasonably assume antibody count indicates immunity. It would therefore be necessary to conduct experiments judging vaccine efficacy based on actual disease prevention instead of humoral response. But this isn't done.

Unfortunately the issues don't end there. Not only do researchers not ask the right questions, most don't even apply the appropriate drug research methodology for vaccine testing for the questions they do ask. That procedure

---

[89] See for example [160].

is known as a *placebo-controlled, randomized double blind*[90] (*pcrdb*) study. Of those who follow it, the vast majority don't employ a true placebo[91].

Moreover, as if these shortcomings weren't enough, I'm not aware of any required vaccines whose research and development effort was not funded by vaccine manufacturers or by those otherwise with a vested interest in vaccines sales.

I would encourage anyone contemplating taking a particular vaccine to seek true *pcrdb* studies (with true placebo) conducted for that specific concoction, where the authors tried to ascertain actual disease prevention and where the research was not funded by parties with financial interests in the sale of the vaccine.

---

[90] The point is to compare a group of people taking a drug, such as a vaccine, to a group *not* taking it (true placebo). If the group taking it is healthier, then there is support for the notion that the drug is safe and effective. If on the other hand we compare one drug to a reference drug (which is typically the case with vaccines), what can we conclude? That depends. For instance, if the reference drug makes people unhealthy, then by comparison it could make the test drug seem like it's good for people when it isn't. Generally we can't conclude from results that a drug is safe and effective if we don't compare it to people who have not taken it. See *Appendix ii* for a full discussion on the placebo controlled double blind study ethos in the context of vaccines.

[91] ibid.

Alas, you'll likely not find a single one free from conflicts of interests nor one testing for disease prevention [114], let alone adhering to the other criteria I describe.

Considering that, and especially because most published research of this type[92] is evidently false[93], it's not only unwise to take what vaccine manufacturers claim at face value through published work, such work should arguably also be inadmissible in the peer-review process for reasons of financial conflicts alone. But unfortunately, the faculty of medical science at large seems to be immune from such ethics. But I digress.

Without the benefit of credible, controlled experiments asking the right questions, we are relegated to uncontrolled ones, also known as *natural experiments*—i.e., observations of situations as they naturally unfold, without the benefit of controlling variables at will. But that doesn't mean we can never draw useful conclusions from natural experiments. It depends on the outcome and the extent to which

---

[92] Generally anything involving statistical theories would fall in this category.

[93] See for example [115], and refer to the phrase *p-hacking*, which is an unwitting research artifact that defeats the ubiquitous employment of the *p-value* paradigm in statistical research. P-value is the standard by which researchers derive confidence in results.

we understand particular circumstances, as will be evident shortly.

Again, the premise of a vaccine is that its administration prevents specific disease if given to a sufficiently high percentage of a population. The presence of disease despite such vaccination unambiguously proves the vaccine false for the targeted disease.

But if we observe no disease emergence it could be one of two things:

(A) The vaccine prevented disease; or,

(B) There was no disease to prevent.

We have no way to determine whether case *A* or case *B* occurred. Thus, whereas the presence of disease following vaccination proves the vaccine didn't work, the absence of disease post-vaccination is clearly inconclusive. So even if modern populations are free of vaccine-targeted diseases, we cannot conclude that this is evidence in support of such vaccines. On the other hand we can conclude certain vaccines as false if we observe no disease prevention despite their introduction.

As I detailed in an earlier passage, there have been multiple, unrelated instances of fully[94] vaccinated populations

---

[94] Meeting the aforementioned herd immunity threshold.

experiencing outbreaks of disease the vaccines were supposed to prevent. The vaccination rates were above those
considered by immunization theory as conferring total protection across the population. [108-113]

Barring incompetencies or execution errors (administration, vaccine preparation, etc.), those total failures are
indisputable proof that the theory that any of those vaccines works is false.

However, it is possible that these could have all been
instances where the vaccine preparations were faulty or the
administration protocols weren't executed properly.
Though the authors of the relevant papers explored those
possibilities and could not conclude such faults, they could
not rule them out either. So here again, the evidence is
strictly inconclusive.

But it should be intuitive that the more instances of
failure the lower the likelihood that they resulted from execution error, unless we know for a fact to expect that kind
of incompetence[95] in the industry—and that's hard to
know. Generally, the more times we don't get what we ex-

---

[95] The fact that the vaccines didn't work is by itself useful information, even if it was due to industry-wide endemically rampant execution error or incompetence, because obviously it would suggest
steering clear of vaccines for that reason alone.

pect theoretically the more likely that it was due to the theory being wrong.

Despite this very intuitive perspective, vaccination orthodoxy apparently never considers the possibility that such vaccines don't work. If proponents of a theory habitually evade contradicting evidence, it becomes less and less likely that the theory is actually falsifiable in practice, even if it might be in principle. Their sheer posturing on the matter renders it immune to disproof.

PARABLE OF THE ERADICATED MALADY

We as a society mostly don't vaccinate adults. It's only required for children to attend school. Therefore we don't have the minimum vaccination rates (70-90%) that would in theory imply eradication. And yet vaccination is widely credited with eradicating disease. How could we have eradication without sufficient vaccination rates? It doesn't add up.

Neither does the historical evidence support the eradication-by-vaccines view. Measles deaths fell sharply well before vaccine introduction[96], for example. And even if decline in deaths had followed introduction, such as appar-

---

[96] See, for example, the measles plot in [123].

ently with polio[97], it would prove nothing, since correlation does not imply causation.

Another fact that does not lend support to the vaccine narrative is that many other diseases sharply declined as well around the same time[98], diseases for which no vaccines existed. To conclude that vaccine-targeted diseases like polio declined necessarily because of vaccination requires fully ignoring the question of why the other diseases declined at the same time without vaccines.[99]

It should be clear that the historical evidence therefore lends no obvious support to the claim that vaccination eradicated any disease.

IMMUNITY FOR IMMUNIZERS

In 1986 Congress passed a law[100] exempting vaccine manufacturers, distributors and healthcare providers (vac-

---

[97] Ignoring possible changes in polio *definition*—i.e., from symptom- to viral strain-centric—which may possibly explain the apparent disease decline. [121][122][124]

[98] e.g., diphtheria, scurvy, tuberculosis, tetanus, pertussis.

[99] A competing hypothesis exists which may better fit observations, which is the idea that sanitation, technology—e.g., refrigeration preventing food spoilage—and nutrition protect against illness by minimizing exposure to toxins. [125]

[100] The National Childhood Vaccine Injury Act [126]

cine administrators) from vaccine related lawsuits. From
that time onward vaccine injury claimants have been di-
rected to the National Vaccine Injury Compensation Program (VICP), more commonly known as the Vaccine
Court.

A per-dose excise tax[101] to manufacturers ultimately
funds payments to those who successfully petition the
Court for injury claims. As a result, vaccine makers enjoy
complete immunity from private lawsuits[102]. This should
undoubtedly give any sane person considerable pause.

Since children by and large must, by law, be vaccinated,
what incentive do manufactures have to make safe vaccines
that work? What incentive would manufacturers have to
publish results or fund research ultimately showing vaccines are not safe or that they don't work[103]? We are to assume that the profit motive, combined with liability exemp-

---

[101] See [157] and Section 5 in [158]. Although the tax is technically
levied from the manufacturer, all of the covered vaccines are U.S.
government-mandated (for children attending school). Manufacturers are free to build the cost of the tax into the sale price, transferring the financial burden onto the obligate consumer.

[102] The United States Supreme Court affirmed in *Bruesewitz v. Wyeth*,
562 U.S. 223 (2011), that vaccine manufacturer liability is "expressly
preempted by the Vaccine Act.'"

[103] Recall that manufactures are essentially the only publishers of
scientific studies on vaccines.

tion, could never possibly compromise the intentions of gigantic, for-profit, international corporations with revolving doors and overwhelming influence on the very legislative bodies which ultimately exempted such corporations from legal product liability in the first place.

If vaccines didn't work or were detrimental, how would consumers know? We've seen above how it's not possible in practice to firmly conclude anything from the available studies and observations as to effectiveness and safety; and, as noted in *Appendix ii*, ill effects are difficult to ascribe to vaccines if they're not specifically expected or if subjects aren't followed long enough.

Couple those limitations with the existence of legislative mandates for kids, and we have a recipe of vaccine maker immunity to market pressure. Manufacturers have no real incentive to make products that work and that don't sicken people.

If the theory of vaccination is at best inconclusive or unfalsifiable, we have no means to determine if the adoption of vaccination schedules has increased or decreased population health.

If the mainstream science on vaccination is untrustworthy it is impossible to logically ascertain whether it is a net benefit or detriment. However, the very existence of VICP and the Vaccine Adverse Event Reporting System [159] in the U.S., a drug safety database dedicated solely to vaccine-caused injury and death, is evidence that harm from vaccines has occurred in considerable number. Those injuries require treatment, much of which undoubtedly result in additional pharmaceutical sales.

Thus even if we suppose that pharmaceutical companies don't profit from the sale of vaccines, they clearly do from vaccine injury. Might this indirect profit from, if not direct sale of, vaccines have been the primary motivation for the industry push to enact VICP? Has vaccine injury occurred at a rate and severity that would not otherwise have occurred without vaccines? Would pharmaceutical sales be lower without vaccines?

Would an industry need to lobby Congress to protect it from liability from the sale of a product having a net benefit to society? These are some of the types of questions we should ask ourselves in the face of government-mandated medical treatments and procedures.

IV. g-Whiz Theory

GIFTED NAÏVETÉ

When I was a teenager, my parents were having marital issues. So as middle class couples with children were wont to do in such circumstances in the 1980s, they opted for a professional therapist.

And because apparently my parents felt that their relationship challenges were affecting me, they asked me to schedule my own regular private sessions with Larry, the psychologist. That's how it was sold to me anyway.

This is where I learned about the concept of general intelligence, or g, and Intelligence Quotient, or IQ [127]. I hadn't really thought about this kind of thing before. One day Larry told me he would ask me some questions to assess something or other about me—I wasn't really paying attention—but he studiously avoided the term "IQ" and never mentioned "intelligence" before administering what would be a very standard and official IQ test, the kind accepted by the likes of MENSA.

I registered as very gifted. However, throughout the oral examination sessions I would have no idea that I was taking an intelligence test. How could a person ostensibly so very bright not realize he was taking a test designed to measure just how bright he was? That irony would dawn on me much later. So much for being very gifted.

If you score very well on an IQ test it says that you're "smarter" than a relatively large number of people, ostensibly. More precisely, the score tells you the proportion of people intellectually "inferior" and "superior" to you, respectively. For he who scores well, it's the narcissist's dream.

But ultimately there was something nagging me about IQ. It posits that mere responses to a finite set of questions or tasks suffices to imply that, barring a statistical margin of error, one person is better, or worse, than many others at literally *everything* the mind can do. It's a rather bold statement.

Because of that, I fully expected to find eventually an official, widely accepted definition for this thing called intelligence.

Curiously though, no IQ test appears to explicitly adopt a concise definition of intelligence, the very quantity it purports to measure. But how can we measure something if we don't know what it is? And yet people presume to measure it [128], and entire careers are based on it.

So why continue further on this topic? We should be satisfied, at least in a technical sense, that without a definition the claim that intelligence is a measurable quantity is not a falsifiable hypothesis. Because how could something undefined admit a test of its own falsity? It can't. But it's worth showing a prominent example of just *how* it can't.

I also believe that we can learn valuable lessons from applying an inchoate idea as if it were complete. We can then gain useful insight from trying to find a falsifiable framework for it. An initially vague and unscientific notion can be an inspiration worth cultivating, as it may mature into something quite the opposite. What's important is to not treat it as if it were scientific before it is.

But first, let's explore a bit about the theory's uses and motivations.

THE INSTITUTION OF *g*

The IQ, or *g*, philosophy of testing mental ability, as opposed to knowledge, has long been adopted in academics, as evidenced by the existence of a wide range of standardized tests—e.g., GED, SAT, ACT for college entrance—which, similarly to IQ, rank people in terms of ultimately a one-dimensional overall score emphasizing generalized ability over acquired skill. This is true to the university graduate level—e.g., GRE, LSAT, MCAT, GMAT.

But the practice of IQ measurement, and its public acceptance, dates back to much earlier. *Buck v. Bell*, for example, was a landmark United States Supreme Court decision upholding the 1924 Virginia Eugenical Sterilization Act allowing for the forced sterilization of inmates with low intelligence based on IQ test results[104]. "Three generations of imbeciles are enough," wrote Justice Oliver Wendell Holmes in 1927. The decision has never been overturned and has led to many tens of thousands of sterilizations by the 1970s. It also bolstered the eugenics movement for a time. [129]

---

[104] Administered through the Stanford-Binet IQ test.

The entire court decision rested on the belief that IQ testing is a valid, scientifically sound practice, and that therefore a universal mental ability called intelligence is a concept which government may legitimately invoke to justify the complete abrogation of the very basic human right of individuals to reproduce.

How does a judge know that the notion of intelligence and its testing methodology is valid? He doesn't. Judges are not usually trained in scientific disciplines or methodologies, nor does the judicial system consider it their role to offer opinions in specialized fields; that responsibility is delegated to so-called expert witnesses.

But what happens if the academic foundations behind expert testimony happen to be flawed or corrupted by corporate or other special interests, for example? That's when we have a big problem.

Does it not make sense for us to develop the ability to judge, as much as we can, the validity of such academic claims? Given what is at stake, is it warranted to place blind faith on the expert testimony of an academician representing the establishment view in a world in which people were for a time prosecuted and executed for witchcraft by this very establishment[105]?

So, in that spirit, let's analyze the intelligence testing philosophy and methodology and try to critique its point of view.

MEASURING THE UNDEFINED

The process of measuring intelligence[106] [131] follows these steps:

1. According to a given psychometric school of thought [132(a-e)], test questions are devised and grouped into c number of putative, cognitive test categories, or dimensions.

2. The administration of all test category questions constitutes an IQ test.

3. Each test category results in its own scalar performance score, producing a c-dimensional point (a set of values with c elements) representing a "raw" overall IQ score—e..g, if there are three categories,

---

[105] [161][162].

[106] The different psychometric methods used to determine a number purported to measure intelligence, such as IQ and *Spearman's g* factor [134][135].

the raw score for one person taking the test is a three-dimensional point, or a set of three values.

4. N number of individuals take the IQ test, the more the better, resulting in N c-dimensional points. For example, a three-category IQ test given to 1,000 people results in 1,000 three-dimensional raw points.

5. The N c-dimensional points mathematically reduce to corresponding scalar (one-dimensional) values by some form of statistical *factor analysis* [133], which looks for a single cognitive variable (typically composite, or "latent") that best explains the differences among test takers. A typical such procedure is to *project*[107] those c-dimensional points onto the axis of

---

[107] For those without sufficient background in or recollection of basic Mathematics, suppose we had an X-Y scatter (two dimensional) plot. Draw a line from one of those points to the X axis, such that the line is perpendicular to the axis. Observe what value that line crosses on the X axis. That number is our "projection", and in this case the projection results in taking a two-dimensional point and reducing it into a one-dimensional point (known as a *scalar*). If we had an X-Y-Z scatter (three-dimensional) plot we could project a point onto a plane, rather than onto a line as above. That's done by drawing a line from the point to, say, the X-Y plane, such that the line is perpendicular to the X-Y plane. That would reduce the dimensionality from three to two dimensions. But in this section we're specifically talking about going from multiple dimensions to a single dimension.

greatest variance by *Principal Component Analysis* (PCA) [136].[108]

6. The resulting scalar values are then rescaled such that the mean has the value 100. At this point we have "IQ scores".

Again, this process assumes no common definition of intelligence, yet a general analytical procedure is well defined for computing intelligence scores. As previously mentioned, without a definition, IQ is not a falsifiable theory.

In effect, IQ suffers from circular reasoning, because it basically says that, "your intelligence is defined by a score which defines your intelligence".

But it's not just a matter of falsifiability. That may be resolved with the eventual adoption of a definition of intelligence, if one is possible. The problem is also that of formulation, as I will explain.

---

[108] The most important thing to understand here is that, because of the procedure, the test takers are in effect being compared in terms of abilities among which they differ the most. Those abilities may be the literal variables (original cognitive categories) being tested or they could be mathematically composite categories (called "latent variables") of the original variables. For example, in a test with three categories—vocabulary, math, spatial reasoning—people may differ the most on some linear superposition of the three instead of, say, on vocabulary alone.

A FORMULATION IN THE OFFING

We humans are living creatures. Therefore any measurable human characteristic—ability, quality, trait—must somehow relate to the act of living in this world. That would include any notion of intelligence, because whatever it is, it's an innate human characteristic, however latent or hidden it may be.

Let's look at the context of humanity to motivate some reasonable hypotheses and perspectives on what intelligence might be. It should be entirely self-evident that we have evolved, by whatever process, into necessarily mutually reliant social creatures for survival: we live in close knit, cooperative societies into a kind of obligate interdependence.

This hypothesized quality of intelligence, is it something which is necessarily a trait of the individual per se or could it possibly be a phenomenon emerging from the collective? Would it make sense to measure an ability, assuming it was defined, at the level of the individual if it happened to be an emergent quality of social groups? It would seem like an important first question to address.

This is an early and critical fork in the road in our quest for a falsifiable theory of intelligence. If we assume that the ability in question is socially emergent, we must reject the notion that testing the individual has any validity. Otherwise we must take the position that testing the group as a whole would be illogical. Which direction we take ultimately determines success or failure; and thus it's critical we get this right.

Does academia's IQ philosophy get this juncture right? Let's try to find out, through illustration.

Suppose we are looking for a way to measure "orchestral intelligence" and we devise a test administered to individual musicians. We all know what orchestral intelligence means: it's how good an orchestra is. We just don't have a precise definition for "good". It remains subjective, just as with intelligence.

Spoiler alert: orchestral intelligence is an emergent property of a group of musicians. It's necessary that people in the orchestra be good musicians individually in their respective instruments, but the way they play together is an ability that a single individual could not possibly possess. The quality we're looking for is a kind of complementarity that enables a group ability to emerge from a collection of individuals specializing in their respective roles.

There may be abilities common to all musicians, such as sense of rhythm, sense of pitch, hand-eye coordination, but these abilities don't address the orchestra holistically— the kinds of targeted abilities that would be consistent with an "IQ" formulation of orchestral intelligence. We instead should think of an orchestra as an entity onto itself—i.e., the whole greater than the sum of its parts. An orchestral personality emerges out of the musical interaction of its members.

Knowing the roles musicians can play in an orchestra, we can in principle devise tests (auditions) to make sure we choose people with the most abilities in those respective roles before putting them together to form an orchestra. And that's what happens in practice: people audition for respective roles—violist, oboist, cellist, pianist, clarinetist, harpist, conductor, saxophonist, etc. They're not each given an identical exam designed to measure what an orchestra can achieve as a whole, whatever such a test might entail.

As reasonable as that approach sounds regarding or- chestras, keep in mind that it can't address the aforemen- tioned complementarity, the quality of working well to- gether. That can only be tested with all the musicians play- ing together. But an outright combinatorial search of the best complementary *and* individually talented musicians is not generally practicable. So the default is usually to just audition people individually. But again, they're not examined by some identical criteria having to do with "orchestral intelligence"; they're necessarily tested according to requirements of individual roles.

So if we carry on with the assumption that the individual posses orchestral ability, we will proceed to no avail, even if we eventually settle on a concise, objective definition of orchestral intelligence.

Now, the IQ camp met with the aforementioned fork in the road many years ago, wittingly or not, assumed that intelligence is per force a property of the individual, and they have not looked back since.

The arbitrary focus on individual over group ability only serves to exacerbate the ambiguity already plaguing IQ theory due to its lack of definition. On the whole it is a pseudoscientific theory, predicated on circular reasoning, dogma masquerading as science. It therefore can make any conclusions its architects desire, without fear of being proved wrong or of losing its appearance of legitimacy. Its ability to evade falsifiability gives its exponents the power to insidiously marginalize or otherwise harm individuals based on pretense. I'm not suggesting that this has been the IQ movement's de facto charter, but that false and pseudo-intellectual status opens the door to great malfeasance, or at least to grave, unintended consequences.

One major historical example of profound misuse, or possibly malfeasance, depending on how you view the eugenics movement, is the aforementioned human rights-abrogating Buck v. Bell decision and its aftermath. That's how important it is to recognize a claim falsely presented as scientific.

V. Seven Criteria

VALID PROPOSITIONS

Suppose the government of country X warns its citizens that foreign country Y has both the means and the deliberate plans to invade. The government of X, faced with a paucity of citizen volunteers, enacts a military draft to carry out a preemptive strike against Y.

If the prediction propounded by government X is accurate, a strike would spare citizens of X the harm of invasion. If the prediction is false, a strike would result in needless loss of life on both sides.

It would behoove the people of nation X to know whether the theory is false. But before they could even attempt that analysis they would need to determine whether the theory is falsifiable.

The substantial challenge, as I've showed in the preceding chapters, is that such real-world claims may sometimes appear falsifiable when they aren't.

While examining various theories in this book under a falsifiability lens, I've induced a set of criteria for what would constitute a valid theory—in the pure or statistical sense.

Thus, in summary, a valid theory is falsifiable and:

I

Unambiguous

Is not ill-posed or vague, is clearly and completely defined and non-circular;

II

Objective

Is not literally or in effect predicated or founded on subjective judgment or absolute decree such as under color of authority, consensus[109], reli-gion, ethics, morality, politics, etc;

III

Predictive

Makes or implies claims which attempt to predict something from an initial state of knowledge and is not merely a narrative in retrospect about observations;

IV

Reproducible

Its essential experimental or methodological pro-cedures are completely defined and openly avail-able such that the identical experiment is at-temptable independently by anyone;

---

[109] Consensus is distinct from repeatability in that the former mere-ly refers to a collective agreement without implying independent conclusion—e.g., the establishment press' notorious appeal to con-sensus on the subject of Climate Change [19] as though scientific conclusions were the product of mere agreement.

V

Non-categorical

Is not explicitly or implicitly a forgone conclusion and does not categorically claim to be indisputable truth;

VI

Simple

Is the simplest among equivalently successful theories (i.e., Occam's Razor)[110]; and,

VII

Bounded

Does not require an infinite or unknown number of tests or amount of time to attempt to prove the theory false, or provides statistical bounds allowing for various degrees of objectively measured confidence that the theory is false.

---

[110] If we're allowed to make a theory as complex as we want, we would be free to devise one so complex that it would be a lookup table of exceptions. Such a theory would be able to "explain" anything we wanted, by adding another table entry, evading falsification.

Most of the criteria and rationales for their inclusion should be self-explanatory. But in what follows I will expound on a few which may require a little more development to grasp their essentialness.

Criterion 3 – A valid theory is predictive

When we have a model of how something works it is a prediction about the future. It's how we can accomplish anything in the real world. You're able to walk because you can predict that dropping your weight forward a bit and then catching it with your foot will result in advancing a certain distance. That's a prediction. You have a model of how walking works, whether you realize it or not, and it is a testable mechanistic theory for which you have much support—no pun intended.

But if you've never walked nor witnessed anyone walking, from your perspective walking would just be an explanation, a narrative about how something has happened. It would be an untested theory. This is not a problem, so long as you could readily test it. But sometimes such explanations aren't testable, either in principle or in practice, as I discuss next.

Criterion 3 – Reproducing Natural Selection

Consider the notion that giraffes have evolved long necks by an instance of the putative Natural Selection process to reach fruit on tall trees. The idea is that giraffes which could not reach fruit died without reproducing their short-necked kin. Consequently, nature selected for those giraffes who could reach fruit, and that's why giraffes have long necks. It's very compelling as an example of a larger Darwinian principle, but the classic Darwinian Theory of Evolution is not actually falsifiable. One reason is that it makes no predictions. Since the theory only makes statements about how things happened in the past, there is no way one could possibly prove it false if it were false.[111]

The Natural Selection notion only gives a narrative about how something has already occurred. What would render Natural Selection strictly falsifiable would be 1) the ability to make a prediction about a specific Natural Selection adaptation before it happens or before it's discovered in the fossil record, proving the theory false if the predic-

---

[111] If we consider specific adaptation claims of Natural Selection, they could in principle be falsifiable if practicable experiments can be devised. See the following section for a discussion around this possibility.

tion doesn't occur (or proving that this one particular prediction was true[112]) and 2) the existence of a finite set of predictions which would render the theory false should those predictions prove wrong[113].

A theory is not necessarily false if it's not falsifiable. It just means we have no ability to test for whether it's false. An unfalsifiable theory is equivalent to a decree—i.e., a declaration by mere authority. So if you're comfortable with diktats, for example, you are free to heed them. My main objective in this book is to, at the least, avail you with the awareness of what you're getting into, epistemologically.

Criterion 6 – Occam's Razor

Why is it important to be in the habit of choosing the simpler of equivalently successful models in a real-world situation? Certainly a theory need not be the simplest to be provable as false (falsifiable). It just needs to admit a test that could render it false. However, obviously the simpler of the two equivalently successful theories is preferable because the more complex the theory the harder it may be

---

[112] Recall that falsifiability does not test for whether a theory is true, just if it's false.

[113] See next section, under *Vaccination, Natural Selection: Criterion 7.*

to falsify it, sometimes convoluted enough to render impractical any theoretically available test of its falsity.

If the government, for instance, attempts to compel you to do something against your will based on a falsifiable theory too impractical to test, the existence of an equivalent but simpler one drawing the same conclusion would obviously be preferable.

In the case where the theory is a mathematical model, the more needlessly complex it is the more likely it's wrong. It's a well known phenomenon called overfitting. The more complex the model, the more it latches on to idiosyncrasies, noise or incompleteness of the observation data, at the expense of prediction accuracy.

In statistical contexts, I argued by example in Chapter III that generally the greater the experimental complexity the higher the uncertainty of its results. Thus an appealing feature of Criterion 6 is evidently also that it naturally selects the formulation requiring the lesser statistical uncertainty (greater confidence) around conclusions, which is obviously preferable.

APPLICATIONS

Theory of Intelligence: Criterion 1

We saw in Chapter IV that the theory of intelligence does not even provide a definition. Instead it starts with prescribing a measurement approach for something undefined. And then the measurement results (*IQ, g-factor*) or their supposed or measured correlates (e.g., financial success, career advancement, formal academic achievement) serve as the de facto definition of intelligence. In so doing, the designers of the test can devise it to correlate arbitrarily with anything they prefer. IQ is therefore the product of circular reasoning. So it fails on the very first criterion.

Climate Change

I showed in Chapter I that "Climate Change" is reducible to one essential claim which is falsifiable in practice[114]—the Greenhouse Effect (GHE)—without requiring expertise in climatology. I even went so far as proving it false. That's certainly enough to discredit the entire Climate Change narrative, but doing so required at least some basic intuition about everyday physics. Most people have that,

---

[114] Test it against the seven criteria.

but not everyone. So here I will evaluate the falsifiability of Climate Change from the perspective of someone having no intuition into everyday physics relevant to the GHE. If we can show that the Climate Change claim fails at least one of the above criteria, it means that it is religion masquerading as science...

Climate Change: Criterion 2

Criterion 2 should be fairly self-evident, except that it may not be immediately apparent to most people that "scientific consensus" is an oxymoron. There is nothing scientific about taking a vote on how nature works. Newtonian gravitation doesn't owe its successes (nor its failures[115]) to a mere popularity contest. We could vote on what happens to a pebble when let go from a height, but the actual answer clearly wouldn't depend on the ballot tally. That's an indisputable example. But what happens when the experiment requires relatively rare expertise to reproduce properly, or what if the evidence is inconclusive? Tallying expert opinion in that case would seem very attractive, but it

---

[115] See Einstein's more experimentally successful, and ultimately more informative, model known as the General Theory of Relativity.

would be just as inappropriate as in the pebble drop experiment.

It's tempting to delegate observation to experts presumably better able to observe such things than ourselves.
But the moment we do so we automatically ascribe authority to them and simultaneously abrogate the faculty of science. Authority is precisely the antithesis of science.

When faced with an inability to make a scientific conclusion about something ourselves, for want of sufficient
expertise, relying entirely on that of others doesn't make it
science. It only accustoms us to conflating science with authority. It is a dangerous practice because, as explained in
Chapter I and developed throughout this book, scientific
reason is the best way we can determine what is true, or as
close as we can get to it. Delegating to only the select few
the power of deciding what is true opens the door to deception, manipulation and other forms of control.

I'm not suggesting that you should dismiss off hand
what you are told by physicians, mechanics, electricians,
physicists, lawyers, but I am saying that the onus is on you
to maintain healthy skepticism and to educate yourself so
you may at least spot swindles or spurious claims. It is an
extensively developed theme of this book that non-experts

are capable of appraising the validity of any claim without bona fide expertise in the applicable subjects.

Climate Change: Criterion 3

What fails "Climate Change" as a whole on Criterion 3 is that evidence contradicting the prediction that the globe is warming doesn't motivate its proponents or its establishment narrators to question the theory or seek alternatives. Instead they offer ex post facto explanations that purport to account for failed predictions. When a theory fails to predict something correctly, proposing mechanisms to explain why we didn't observe what the theory predicted is a way to avoid testing the theory. Such proposals are not themselves predictions, and thus not falsifiable, which implies that they are not scientific.

For instance, when winters in widespread parts of the Northern Hemisphere, especially North America in 2010[116], became unusually cold (and wet) when the global average temperatures had been flat since 1998, the explanation was that "Climate Change" has the effect of destabi-

---

[116] An internet search for "polar vortex 2010" will bring up relevant sources, including this Wikipedia article (as of this writing): https://en.wikipedia.org/wiki/Polar_vortex

lizing the polar vortex, which causes unusually cold air to descend into more southern latitudes from the pole [137].

If the unusually cold winters had been predicted by the vortex theory, it would have possibly[117] qualified as one element of support for the Global Warming / Climate Change theory. But the vortex phenomenon was proposed *after* the anomalous winters had arrived, as explanation for what had already occurred. Therefore it was not a prediction. At the risk of beating a dead horse, let me emphasize that there is no way to test that a prediction comes true if the "prediction" comes after the fact.

And after the years 1998-2013 saw no global warming as gleaned from publicly available data, exponents of Climate Change proposed several explanations, including that the oceans had absorbed the missing heat, as reported by

---

[117] It would certainly not detract support. But ultimately, determining causality is not a simple matter. Ideally, controlled experiments in, say, smaller scales would be more definitive. But the point is that proponents of the vortex connection didn't declare it until after the fact. So causality is a moot point.

Science Daily[118], and that the data measurements underestimated the actual temperatures[119].

That the oceans absorbed the heat [130] is a plausible explanation, but again, not a falsifiable hypothesis. It is merely a narrative about something which occurred in the past, not a formulation describing a prediction. And the idea that the data underestimated the actual temperatures originates as mere speculation[120,121], yet it was headlined as definitive explanation in the preeminent scientific press[122,123].

------

[118] "Oceans act as a 'heat sink': No global warming 'hiatus'," Science Daily, November 22, 2016. https://www.sciencedaily.com/releases/2016/11/161122182458.htm

[119] [130][138-140].

[120] "Incomplete global coverage is a *potential* source of bias in global temperature reconstructions," (emphasis mine) [140]. "Climate models projected stronger warming over the past 15 years than has been seen in observations. Conspiring factors of errors in volcanic and solar inputs, representations of aerosols, and El Niño evolution, *may* explain most of the discrepancy," (emphasis mine) [141].

[121] [138][140][141].

[122] "Climate-change 'hiatus' disappears with new data" [139]. Despite that assertive headline, the article, which cites [138][140][141], admits that "An apparent pause in global warming *might* have been a temporary mirage," (emphasis mine).

[123] [139].

Climate Change: Criterion 5

As anyone knows from the mainstream press, the phrase, "The Science is Settled" has been ubiquitous, having the effect of preempting debate on the subject. In other words, no dissent dissuades the Climate Change authorities from promoting it as being true. That clearly renders Climate Change an a priori, categorical, or foregone, conclusion. If this seems very obvious to you, congratulations, because it's apparently non-obvious to the bulk of the populace around the globe.

Therefore, in conclusion, the Climate Change narrative as a whole fails the Seven Criteria test without our being experts in any kind of climate science. We can see that no objectively contradictory observation is ever admitted as evidence that the theory might be wrong. No counter-evidence appears to change the stance that the planet is certainly doomed.

This brings me to the notion of *emergency trap*. It's the better-safe-than-sorry argument I examined in Chapter II. It says that we don't have the luxury to wait and see what happens; we must instead act now to avoid a presupposed catastrophe. But as I expounded in Chapter II, that argument is an easily discredited logical fallacy, because it takes no account of the costs associated with acting versus not acting to address the ostensible emergency. The key lies in evaluating first if the emergency is real, but this is what is precisely and studiously avoided by promoters of such a trap, which should at the very least give you pause in and of itself.

Vaccination, Natural Selection: Criterion 7

The principles of vaccination and Natural Selection, respectively, are examples of *unbounded* theories. I've already expounded on vaccination with regard to Criterion 7 back in Chapter III, so I will discuss Natural Selection here.

First, as I've argued before, Natural Selection as a whole does not predict anything (fails Criterion 3) because it merely provides a narrative framework for proposing explanatory mechanisms in *hindsight* for specific observation of what has already occurred. But suppose we claim to predict a specific Natural Selection adaptation which has not already occurred or has yet to be discovered in fossils, rather than merely providing a hindsight narrative. That could render that *instance* of Natural Selection theory falsifiable.

For example, say we create a confined habitat comprising natural prey representing fur of all colors amid a totally white environment. We call this the *test* group. The hypothesis is that the test group prey will adapt fur color to be preferentially[124] white if we introduce predators to its environment.

We reason that the white coats are harder to contrast against the white background, conferring protection from premature death by predation. White prey in the test group will thus reproduce more than prey of other colors, the theory goes, resulting in a preferential selection for white-furred descendants.

For comparison, we set up the same experiment with a different group of prey but with a habitat background representing a range of natural colors not contrasting with any particular fur. We call this the *control* group.

Suppose we find after ten generations that the white fur prevalence in the test group is not significantly greater than that in the control group. So we wait another ten genera-

---

[124] "Preferentially" is an arbitrary threshold, but any prevalence threshold will do if we want to satisfy Criterion 7, so long as we set one and stick to it. For example, we could define "preferentially white coats" to be whatever percent dominance resulting from, say, 99.9% confidence that it's not attributable to chance.

tions, but suppose we still don't observe a significant difference. At what point do we decide that our prediction is false? It certainly remains inconclusive, since we don't know if the prediction will come true in, say, another ten or more generations or whether an equivalent statistical measure comes true if we conduct a multiple number of such longitudinal experiments.

Thus in practice we may at best only have varying degrees of confidence that the white fur adaptation theory is false: the more experiments (longitudinal or parallel generations) the more confidence—never 100% proof. If these experiments are practicable—i.e., can be completed during one's lifetime or coherently carried across experimenter generations—it would render white fur adaptation theory falsifiable.

However, white fur adaptation theory, true or false, can only speak to that particular instance of Natural Selection theory, not to Natural Selection as a whole.

Similarly, if we'd predicted the fossil record to show an adaptation not previously discovered, its absence in spatiotemporally finite excavations is not proof of its non-existence. We can't anticipate whether we'll eventually find it (recall the streetlight effect). Therefore we can't potentially prove that adaptation strictly false unless we can account for every inch of the Earth's crust. We would in practice only have some degree of confidence as to falsity. And again, we would still not be able to conclude anything as to the falsity of the entire theory of Natural Selection, neither statistically nor strictly, since one false adaptation prediction does not imply the falsity of the *mechanism* of Natural Selection.

Absent a testable mechanism capturing the Natural Selection process in one well swoop in all its forms, such as an explicitly genetic formulation expressed in closed mathematical terms, proving Natural Selection false requires performing an infinite number of experiments addressing all the possible adaptations it implicitly proposes. And that in itself renders it unfalsifiable[125].

------

[125] I'm here only considering the classical, biological Natural Selection, not its computer science analogue, known as the Genetic Algorithm (GA), which is described mathematically. If the GA turns out to be falsifiable, Natural Selection is not necessarily falsifiable by extension. That would depend on how faithful the mapping is between the two. In any case, the matter is rather complex and beyond the scope of this book.

VI. Going Flat Out

TESTING WHAT MAY SEEM DETESTABLE

Imagine a wide building, preferably circular. Climb to the roof, and stand in the middle of it. Spin slowly around through all 360 degrees while observing the edge of the building directly in front of you.

Notice that the edge is at the same visual height regardless of your orientation.

Move closer to the edge on one side of the roof. Look straight at the edge. Its appearance now is lower. You have to look down a little to focus on it. If at this point you turn around completely (180 degrees) you'll notice the other horizon is higher in your visual field.

Now imagine yourself on a big ship in the middle of the ocean, in calm waters, with no land in sight[126]. In a *flat* Earth, accounting for mountains and hills, if you spin slowly around the way you did when you were on the roof of the building you'd see the same effect: the edge—called the *horizon* on Earth—should be visually at the same height, but only if the ship happens to be at, or sufficiently near, the center of Earth.

So if you sail through the world in that ship in a *flat* Earth, eventually you'll notice that the horizon on one side is taller than the horizon on the opposite side.

But in practice, in your travels, have you ever noticed such horizon lopsidedness? If the Earth were flat, someone surely would have observed this and reported it. But to my knowledge no one has. Of course, it also begs the question, why hasn't anyone witnessed the edge of the Earth up close?

But that may be hearsay, strictly speaking. It's important to observe things ourselves. So we could take a cruise ship around the world or fly to far destinations and see if at

---

[126] This is so we can avoid hills, mountains, valleys, to ensure that nothing's in the way of the horizon or edge of Earth.

some point we observe horizon lopsidedness. For what it's worth, in all my travels[127] I've never noticed it.

If instead we posit that the Earth is approximately *spherical* it would geometrically imply a specific distance between the horizon and the eyes of someone observing it. A six-foot tall person standing on the surface of a sphere the size of Earth would measure the horizon to be about three miles away (closer for a shorter person), regardless of his or her specific coordinate location on the surface[128].

That's very testable. To test this on land, go to an expanse having approximately constant elevation[129] on a clear day, plant something like a flag pole spanning about your height on some spot on the ground, then walk three miles straight away from the pole. Approximately the entire pole should be visible from where you stand, though you might need binoculars to see it well.

As you move backwards further away from the pole, the pole should gradually appear to sink behind the horizon, save for local variations in elevation.

---

[127] I've lived in two different continents, Europe and North America, and I've traveled as far as Japan.

[128] See *Appendix iii*.

[129] Such as in the Great Plains in the United States.

If you're at the beach on a clear day, you could ask a friend to set sail out to sea. If you're eyes are six feet above the ground, when the ship is about three miles out it should appear to "sink" and eventually fall behind the horizon as the ship moves further away from shore. Again, binoculars may be necessary.

As many people know, another way to test this is to observe a tall sail ship, or any tall ship, as it moves away. If the Earth is basically spherical, the sail ship will start to gradually descend behind the horizon when it reaches about three miles from shore, assuming you're standing on shore with not much greater elevation than the sea surface[130].

A PRIEST IN LAB COAT OR SCIENTIST IN CASSOCK?

Generally, that's how science is conducted; we induce a falsifiable theory that implies one or more testable predictions. If any of its predictions prove false, the theory is false. If all prove true, the theory's falsity is not demonstrated[131], but it doesn't follow that the theory is true.

---

[130] The higher up you are, the farther the horizon will appear—see the *H* parameter in *Appendix iii*.

[131] This is what scientists mean when they say that a theory has "support".

The two theories discussed here concerning Earth's shape are both clearly falsifiable. But one is necessarily false, because the Earth cannot be flat and spherical at the same time.

Ironically, many consider Flat Earth theory to be completely pseudoscientific and, say, Natural Selection to be fully scientific. This may be true in terms of the motivations of its proponents, but at least in terms of falsifiability, as I've demonstrated in this book, actually it's the opposite; Natural Selection, unlike Flat Earth, is in fact pseudoscientific[132] in its construction, mainly because it makes no predictions at all, let alone testable ones, whereas Flat Earth theory does.

This in an of itself implies nothing about the ultimate truth or falsity of either Flat Earth or Natural Selection. As usual, falsity depends on the outcome of observations and experiments, whereas a theory's truth is not knowable (at least not by the falsifiability test).

In my observations, and based on the points made in this chapter, I personally conclude that Flat Earth theory is false—but you be the judge, since you *can* test it for yourself. The bottom line is that, like it or not, Flat Earth theo-

---

[132] See Chapter V.

ry is at least formulated in a way that renders it scientific (falsifiable), whereas we could never test whether or not Natural Selection is false in its current form, as I've previously explained. Therefore, unless Flat Earth proponents as a rule simply refuse to ignore any conclusive evidence against it, Natural Selection is actually the religious one of the two.

Epilogue

In the mind's I are the philosopher and the technician. The philosopher asks why, the technician how. Disease begins where the balance between the two ends.

Balance is chaotically dynamic, like negotiating a pole from falling when holding it vertically at its bottom by one finger. The natural laws governing the pole's tendencies are naked simple. And yet it took the complexity of the human brain to enable a mind to uncover them.

In the same poetry, it takes unfathomable complexity to coax the biologically diseased into acquiescence. When flesh dies it invites bacteria into a complex orchestration of emergent processes, countless living things and dances one could never behold in one's mind at once.

Bacteria don't wait merely until the body is drained of life. They assist death wherever assistance of life is not possible. They benefit from the balance of death and life. The body sheds to make room for the newly born, until eventually the balance is lost, and then only certain types of bacteria dominate. It is mere negotiation of *time* before that inevitable end, but time is everything to a living thing.

The most pernicious of microorganisms breed best when the balance of life and death is lost. What is beautiful to the microorganism at this stage is ugly to the would-be carcass. But it is neither. It just is.

As in the biological cycle of life, the degradation of the mind's inner balance is breeding ground for tyranny, the terminal illness of the mind.

If the philosopher takes over, the mind loses sight of *praxis*. It becomes a prodigal demon, victim to its own false realities for lack of earnest orientation from Mother Nature, its obligate partner. It is like a dream out of control, a child born with no sensation nor proprioception. It cannot learn, because it cannot make mistakes. It is it's own incubus and creates its own terminal illness in a cacophony of self delusion and self demolition, dragging all technicians down with it.

When the technician dominates, the mind has no appreciation for *theoria*. It is like a lifeless, yet perfectly functioning body. It therefore craves external control, like a machine lacking executive function, needing its programmer. It is lost without its master.

It thusly invites the tyrant, like the summoning of nefarious gut flora by befallen inner flesh. The process is not binary. It is gradual and may be overturned, but the further progressed the less likely reversed. Unlike the mind, the body always meets the same eventual end. All that is negotiable is time, whereas the mind may live out all its days without tyranny if balance between philosopher and technician is perpetually and diligently sought.

I hope this book aided you in cultivating both the philosopher and the technician within.

Appendix i
Hot Air in the Greenhouse

What follows is a more formal exposition demonstrating the falsity of the Greenhouse Effect, as an alternative to the common sense approach I developed in Chapter I.

As in that chapter, I assume here the same mainstream Greenhouse Effect definition [21(a-d)] widely adopted by universities, scholastic curricula, governing bodies and other influential or well established institutions, such as the IPCC[133].

The Greenhouse Effect is a critical and indispensable premise of Anthropogenic Global Warming theory (also known as "Climate Change"). Therefore, if the Greenhouse Effect is false, so too is Anthropogenic Global Warming theory as defined by the IPCC [1].

Before I begin, let's recall the definitions provided in Chapter I.

---

[133] Intergovernmental Panel on Climate Change [1].

Anthropogenic Global Warming (AGW) Theory

The Earth is warming too much from the Greenhouse Gas Effect (GHE) due to an excess in anthropogenic Greenhouse Gasses (GHGs), principally CO_2. [1]

The Greenhouse Effect [21(a-d)] states:

1. The sun's rays hit the Earth's surface, raising its temperature.

2. That heated surface then radiates energy upward in a band of wavelengths that GHGs in the atmosphere can absorb.

3. The temperature of the GHGs consequently increases, resulting in re-radiation of part of that energy back down to the surface (the rest out into space).

4. The addition of this "back-radiation" intensity from the GHGs to that from Earth's surface causes the temperature at the surface to be higher than it would be without (or with less) GHGs in the atmosphere.

To compute the increased temperature due to the GHE, AGW theorists employ the so-called Stefan-Boltzmann equation (SBE) as follows:

The radiant heat flux (intensity)—let's call that J_E—from the Earth's surface and the radiant heat flux coming back down from the GHGs—let's call that J_{GHG}—are added together and entered into the SBE.

The SBE has this form:

$$J = \sigma\, T^4,$$

Where σ [134] is ostensibly a known, overall physical constant in this context, T is the temperature of the radiating body (e.g., the Earth's surface, the GHGs) and J is the heat energy flux (typically expressed in units of W / m^2, or "Watts per square meter").

If we take the SBE at face value, it would seem that we could solve for T whenever we have some value for J. Indeed, warmists do exactly that.

With the total flux in the GHE case being $J = J_E + J_{GHG}$, the GHE theory states that the temperature at the surface of Earth is calculated this way:

---

[134] GHE theory uses an additional constant, the emissivity, but I absorb that into σ, because that detail bears no consequence to this analysis.

$$J_E + J_{GHG} = \sigma\, T^4.$$

Solving for *T* gives,

$$T = (\, (J_E + J_{GHG})\, /\, \sigma\,)^{1/4},$$

Notice that removing the J_{GHG} component (the alleged GHG "back radiation") makes *T* a smaller number. Hence the claim is that without GHGs we would have a cooler Earth surface, and therefore with the addition of excess amounts of GHGs like CO_2—bigger J_{GHG} term—we would have an even warmer surface. Hence "global warming" due to CO_2. At first blush, this sounds like a pretty reasonable model.[135]

But in calculating the GHE, are warmists correctly employing the SBE?

---

[135] I will not challenge the details of how GHE theorists determine the values σ, J_E and J_{GHG}. I will give GHE the benefit of the doubt in that regard, because, as you will see, their veracity or falsity is ultimately immaterial to the conclusions I will be able to draw. Also, although textbook computations of the GHE use different symbols and slight variations in form, they all fundamentally do the same thing: they add fluxes together to deduce temperature.

Warmists add all radiant heat fluxes (intensities) together and plug them into *J;* then they solve for *T*. That's it. So let's apply that procedure to a thought experiment.

Suppose we have two stones (or any two things), identical in every way, for simplicity. We put the stones inside an oven until they reach the same temperature, *t*.

Immediately after removing them from the oven, we measure the radiant heat energy flux (intensity) emanating from them and find it to be the same from each stone, as we would expect from two bodies at the same temperature.

If we let *J* be the flux from each stone, *t* be the temperature of either stone alone and *T* the temperature resulting from two stones together, applying the SBE the same way warmists do gives:

$$J = \sigma t^4$$

(one stone radiant flux measured by itself)

and

$$2J = \sigma T^4$$

(two stone radiant fluxes together).

We can then compare what we get for *t* to what we get for *T* and see if the result matches what we observe in reality. But what *is* the reality in this scenario? From experience we know that the temperature of each stone individually should be the same as the temperature of the stones together. In other words, the GHE approach should yield *t* = *T*. Is that what we get?

Let's solve the above two relations for *t* and *T*, respectively, which gives:

$$t = (J / \sigma)^{1/4}$$

and

$$T = (2 J / \sigma)^{1/4}.$$

From simple algebra, we get:

$$T = 2^{1/4} t.$$

As you can see, it says that $T \neq t$, which is the opposite of what we observe in reality. That means that we have just proved GHE false. Clearly we now can say definitively that the SBE can't be used to solve for temperature from intensity. It simply doesn't work.

The precise reason why SBE can't be used to solve for temperature despite appearances has to do with the historical fact that the SBE derives from a physical model known as the Planck Law, which is based on principles of quantum statistical mechanics—this is way beyond the scope of this book, but a treatise can be found typically in undergraduate introductory texts of Thermodynamics and Statistical Mechanics[136].

The Planck Law is a model describing the heat radiation that physical bodies emanate due to temperature. It fundamentally says that the color, or color spectrum, of light in that radiation—not necessarily visible—emanating from an object directly relates to the temperature of that object. More precisely it says that the relative quantities of colors (shape of the curve) uniquely determine the temperature of an object. This is fundamentally why astrophysicists measure the light spectrum from stars to deduce temperature.

Put another way, the Plank Law says that temperature is about color (of light), not intensity (radiant heat flux),

---

[136] See for example [25].

whereas the SBE, if taken at face value, suggests that flux (heat intensity) relates to temperature. That's simply not true, and we've just proved that by trying to apply it as though it were.

If you're curious[137]...

The Plank Law is the result of modeling the thought experiment of an imaginary spherical box containing photons. It turns out that the result of such a model in large part correctly predicts the radiation observed from objects due to temperature. Because the thought experiment involves photons, the model adopts quantum statistical mechanics, the subfield of physics which has had the most success to date in predicting phenomena where a collection of subatomic particles is concerned. The Planck Law works very well for solids, but not generally for gasses, incidentally.

As you may have noticed, the GHE applies the SBE, which derives from the Planck Law, even though gasses are

---

[137] Refer to ch. 26 of [25] for more detail, or consult any undergraduate physics text of statistical mechanics covering the Plank Law and blackbody radiation.

involved. But, as you've seen, it wasn't necessary to allude to this fact in order to disprove the GHE.

Technically, the SBE is the "area under the curve" of the Planck Law. The Planck Law curve is a spectrum, meaning intensity versus wavelength or frequency (in this case it's not pure intensity, it's intensity normalized by the color of light). When we take the area under that curve what happens is that we now have a scalar value, not a function. In other words, the spectral character is gone. We just have a single number representing the total energy contribution from all colors. Because the SBE is derived by this process, it has no color information.

The SBE is only a comparison tool. It's not intended to be used at face value. It's meant to tell us that flux depends on temperature to the fourth power so that we can compare the rate of energy loss between *identical* objects at different temperatures—e.g., doubling the temperature results in $2^4 = 16$ times the rate of energy loss of a body. It's not intended as a means to solve for temperature given radiant energy flux. The fact that the SBE seems to allow us to solve for T algebraically is misleading. It's always important to understand the original context of an equation before using it. But clearly, GHE theorists don't understand this point or have overlooked it.

Even if you didn't quite understand any of the physics I just presented, you can intuitively see why the SBE isn't meant to be used to solve for temperature because, from common experience, as mentioned in the previous section, temperature is obviously about color. A blue flame is *always* hotter (has higher temperature) than a yellow one. We can reduce the size of the blue flame so that it has the same radiant energy flux as the yellow flame, meaning that T and J in the SBE do not have a one-to-one correspondence, contrary to what the literal form of the equation would suggest.

But what about all the talk of glaciers melting? What about the polar bears, aren't they drowning for lack of icebergs to land on? Let's suppose for the sake of argument that there has indeed been a trend whereby glaciers have been on the decline for decades and icebergs have rarefied in number because they've melted into the seas due to rising global temperatures. If the theory of GHE is false, as I've shown, can GHE be the cause of such melting? Logically, no.

Ice thinning, glaciers melting, temperatures rising over time, if true, are nothing more than trends in the absence a falsifiable theory which can correctly predict their existence and occurrence. A trend is not a (predictive) model. It's just an observation or story about what happened in the past. Merely observing glaciers thinning over time says nothing about what's going to happen in the future. If a trend were a predictive model, I have some NYSE stock to sell you.

Final Thoughts

In the above mathematical treatment, I gave Anthropogenic Global Warming theorists the benefit of the doubt and proceeded to use the Stefan-Boltzmann equation in exactly their fashion. This led to results not borne out by common experience.

Now I will demonstrate that the theory fails on a more basic level.

In plain english, the SBE says that the electromagnetic radiation emanating from a body having temperature t is proportional to t^4. Let's take a moment here to underscore that, by the equation's very constitution, the body in question (at temperature t) must be the source of radiation. Indeed, its temperature is the reason the body radiates at all.

This is the entire premise of the SBE. Therefore the quantity *t* in the SBE cannot possibly refer to the temperature of any *target* of the radiation.

To put it more explicitly, the equation literally reads like this...

$$J_{source} = \sigma\, t^4_{source}$$

The fundamental problem with GHE theory is that it seeks the temperature of the *target* of radiation (namely the atmospheric region just above Earth's surface). It proceeds as though the SBE says this...

$$J_{source} = \sigma\, t^4_{target} \text{ (incorrect!)}$$

This elementary mistake renders GHE theory a non-starter, and therefore bunk.

Appendix ii
The Randomized, Placebo-Controlled Double-Blind Trial Ethos

The following is a simplified description of the standard scientific method in medical sciences for determining safety and effectiveness (efficacy) of any drug, here in the context of vaccines.

We start by taking a randomly selected set of people from the larger population in society. To half of the subjects we inject a vaccine targeting a specific disease, and to the other set we inject saline (placebo). We call the latter a "placebo control" group and the former the "test" group.

We expose[138] both groups (animals) to the disease which the vaccine is supposed to prevent.

To avoid confirmation bias, we make sure that the experimenter isn't aware which subject got the vaccine and which got the placebo during administration. So at this point we have what is known as a "randomized, placebo-controlled double blind study" experimental

---

[138] As mentioned in Chapter III, vaccine researchers don't do this part; they merely ascertain the presence of a humoral response (antibody production) and assume that's equivalent to disease protection.

design—"double blind" because neither the subject nor the experimenter knows what is in the syringe.

Next we compare the health status of the two groups. If the test group is healthier by a statistically significant margin and to a degree which would justify the cost-benefit of administering the vaccine to the broader population, we market the vaccine.

The process of comparing the health status of the two groups is arbitrary. We have to decide what health conditions to check, or which unintended side-effects (aka, adverse events) to look for. In practice, this logically includes expected side-effects, but certainly not unexpected ones.

For example, if we don't happen to suspect that the vaccine could possibly cause a particular type of cancer, we wouldn't test for it and therefore wouldn't note cancer occurrence as part of a study on the vaccine's safety record.

Another arbitrary parameter is latency. How long do we observe the study subjects for adverse events, a week, a month, years? It's impossible to know that without knowing what could possibly go wrong and how long those developments take to unfold for particular kinds side effects.

It turns out that observation periods for vaccines on the market are on the order of several days[139]. Is that a reasonable amount of time?

For a vaccine observation period of 5 days, as a typical example, that would mean that if on day 6 a subject develops his only serious adverse reaction to the vaccine—say, lifelong autoimmune disease—the experimenter would have recorded him as having had no serious adverse events from the vaccine. Is that an accurate recordation? Is that a reasonable overall approach to assessing vaccine safety?

---

[139] For example, study subjects for the hepatitis b vaccine (trade name RECOMBIVAX HB) "were monitored for 5 days". [142]

Appendix iii

Looking to the Horizon

What follows is a description of how to determine the distance between one's eye and a point on the horizon, assuming an approximately spherical Earth.

Referring to the diagram below, where c is the center of Earth, from the vantage of point e of a person's eye at height H standing on the Earth's surface of radius R at point h, the line of sight from the observed horizon point p to that person's eye at e is precisely the tangent line T of

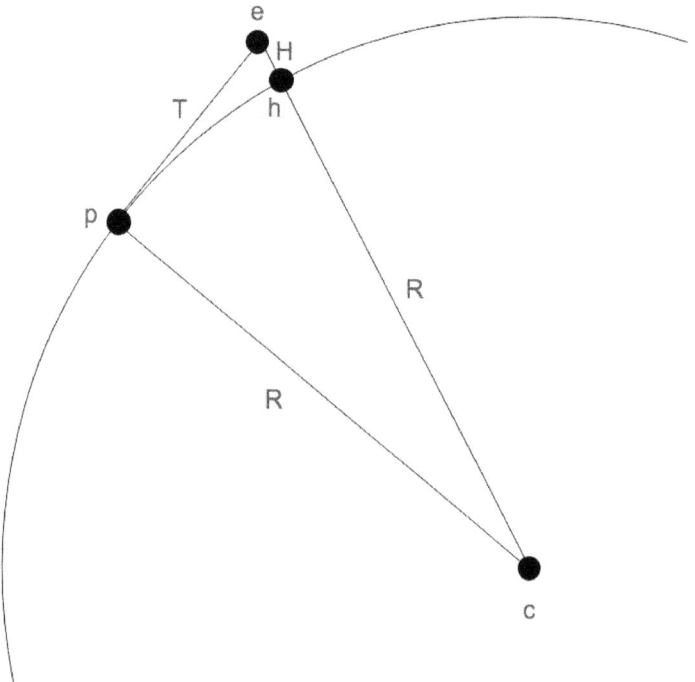

the great circle bisecting the Earth on whose circumference *p* and *h* are located.

Because line *T* is tangent to the circle, the angle between *T* and *R* is 90 degrees, which means that we have a right triangle. Right triangles obey the Pythagorean theorem. And thus we have,

$$(R + H)^2 = R^2 + T^2 .$$

We are interested in solving for the distance, *T*, between the eye and the horizon. Therefore, a little algebra gives,

$$T = ((R + H)^2 - R^2)^{1/2}.$$

The radius of a spherical Earth was evidently deduced long ago[140] from simple measurements to within about 6% of what we calculate today with more modern techniques, which turns out to be about 3,960 miles. So for a six-

---

[140] According to American Physical Society News [143], third century B.C.E. Greek polymath Eratosthenes of Cyrene was able to approximate the circumference of the Earth (which is proportional to 2π times Earth's radius) based on the distance from Alexandria to Syene and angles of sun shadows cast in those respective cities.

foot[141] tall person, the distance from his/her eye to the horizon, is approximately,

$$T \approx ((3960 + 6/5280)^2 - 3960^2)^{1/2} \approx 3 \text{ miles}.$$

---

[141] 6 feet is 6/5280 miles.

Bibliography

[1] Intergovernmental Panel on Climate Change (IPCC). www.ipcc.ch/.

[2] Encyclopedia Britannica. "Pollution | Definition, History, & Facts." Accessed July 5, 2020. https://www.britannica.com/science/pollution-environment.

[3] Wright, Laurence A., Simon Kemp, and Ian Williams. "'Carbon Footprinting': Towards a Universally Accepted Definition." Carbon Management 2, no. 1 (February 1, 2011): 61–72. https://doi.org/10.4155/cmt.10.39.

[4] Kolbert, Elizabeth. (March 25, 2020) "Why We Won't Avoid a Climate Catastrophe." National Geographic Magazine. https://www.nationalgeographic.com/magazine/2020/04/why-we-wont-avoid-a-climate-catastrophe-feature/.

[5] Bracket, Ron. (July 9, 2020) "World's Annual Temperature Could Hit 2.7-Degree-Rise Threshold Within Next Five Years, WMO Says." The Weather Channel. Accessed July 14, 2020. https://weather.com/science/environment/news/2020-07-09-annual-global-temperature-dangerous-rise-wmo.

[6] Tharoor, Ishaan. "Analysis | The World's Climate Catastrophe Worsens amid the Pandemic." Washington Post, June 28, 2020. https://www.washingtonpost.com/world/2020/06/29/worlds-climate-catastrophe-worsens-amid-pandemic/.

[7] Sengupta, Somini. "'Bleak' U.N. Report on a Planet in Peril Looms Over New Climate Talks." The New York Times, November 26, 2019, sec. Climate. https://www.nytimes.com/2019/11/26/climate/greenhouse-gas-emissions-carbon.html.

[8] Paris Agreement to the United Nations Framework Convention on Climate Change, Dec. 12, 2015, T.I.A.S. No. 16-1104.

[9] Kyoto Protocol to the United Nations Framework Convention on Climate Change, Dec. 10, 1997, 2303 U.N.T.S. 162.

[10] Bajželj, Bojana, Keith Richards, Julian Allwood, Pete Smith, John Dennis, Elizabeth Curmi, and Christopher Gilligan. "The Importance of Food Demand Management for Climate Mitigation," August 31, 2014. https://doi.org/10/245933.

[11] "Special Report on Climate Change and Land — IPCC Site." Accessed July 13, 2020. p. 479. https://www.ipcc.ch/srccl/ , https://www.ipcc.ch/site/assets/uploads/sites/4/2020/02/SRCCL-Chapter-5.pdf.

[12] Schlossberg, Tatiana. "Flying Is Bad for the Planet. You Can Help Make It Better." The New York Times, July 27, 2017, sec. Climate. https://www.nytimes.com/2017/07/27/climate/airplane-pollution-global-warming.html.

[13] "05-1120 Massachusetts v. EPA (4/2/07)," 2007, 66. www.supremecourt.gov/opinions/06pdf/05-1120.pdf.

[14] "Massachusetts v. Environmental Protection Agency." In Wikipedia, May 31, 2020. https://en.wikipedia.org/w/index.php?title=Massachusetts_v._Environmental_Protection_Agency&oldid=960043978.

[15] National Geographic. "Air Pollution Causes, Effects, and Solutions," February 4, 2019. https://www.nationalgeographic.com/environment/global-warming/pollution/.

[16] "FROM SEE-SAW TO WAGON WHEEL," BBC Trust. 2007. p. 40. http://downloads.bbc.co.uk/bbctrust/assets/files/pdf/review_report_research/impartiality_21century/report.pdf.

[17] "BBC Orders Impartiality Follow-Up." BBC News, August 2, 2012, sec. Entertainment & Arts. https://www.bbc.com/news/entertainment-arts-19091530.

[18] Carbon Brief. "Exclusive: BBC Issues Internal Guidance on How to Report Climate Change," September 7, 2018. https://www.carbonbrief.org/exclusive-bbc-issues-internal-guidance-on-how-to-report-climate-change.

[19] "The 'mini Ice Age' Hoopla Is a Giant Failure of Science Communication." The Conversation, PHYS ORG. July 24, 2015. https://phys.org/news/2015-07-mini-ice-age-hoopla-giant.html.

[20a] Popper, Karl (1935). Logik der Forschung. Vienna, Austria: Verlag von Julius Springer.

[20b] Popper, Karl (1959). The Logic of Scientific Discovery. London, UK: Hutchinson & Co.

[20c] Popper, Karl (2005). The Logic of Scientific Discovery. (n.p.): Taylor & Francis.

[21a] Goody, Richard (1995). Principles of Atmospheric Physics and Chemistry. UK: Oxford University Press. pp. 142-143.

[21b] "Idealized Greenhouse Model." In Wikipedia, December 31, 2019. https://en.wikipedia.org/w/index.php?title=Idealized_greenhouse_model&oldid=933287736.

[21c] "CHAPTER 7. THE GREENHOUSE EFFECT." Section 7.3.2. http://acmg.seas.harvard.edu/people/faculty/djj/book/bookchap7.html.

[21d] "What Is the Greenhouse Effect?," n.d., 3. IPCC: https://aamboceanservice.blob.core.windows.net/oceanservice-prod/education/pd/climate/factsheets/whatgreenhouse.pdf.

[22] IPCC, 2007: Climate Change 2007: The Physical Science Basis. Contribution of Working Group I to the Fourth Assessment Report of the Intergovernmental Panel on Climate Change [Solomon, S., D. Qin, M. Manning, Z. Chen, M. Marquis, K.B. Averyt, M.Tignor and H.L. Miller (eds.)]. Cambridge University Press, Cambridge, United Kingdom and New York, NY, USA.

[23a] "Astronomy 122 - Measuring the Stars." Accessed September 13, 2020. https://pages.uoregon.edu/jimbrau/ astr122/Notes/Chapter17.html#temp.

[23b] "6.3 Visible-Light Detectors and Instruments - Astronomy | OpenStax." Accessed September 13, 2020. https:// openstax.org/books/astronomy/pages/6-3-visible-light-detectors-and-instruments.

[24] "Thermopile." In Wikipedia, June 29, 2020. https:// en.wikipedia.org/w/index.php? title=Thermopile&oldid=965068054.

[25] Stowe, Keith (1984). Introduction to Statistical Mechanics and Thermodynamics. New York, John Wiley & Sons, Inc.

[26] "WHO Director-General's Opening Remarks at the Media Briefing on COVID-19 - 3 March 2020." Accessed July 9, 2020. https://www.who.int/dg/speeches/detail/who-director-general-s-opening-remarks-at-the-media-briefing-on-covid-19---3-march-2020.

[27] Ioannidis, John. "The Infection Fatality Rate of COVID-19 Inferred from Seroprevalence Data." MedRxiv, June 8, 2020, 2020.05.13.20101253. https://doi.org/ 10.1101/2020.05.13.20101253.

[28] CDC. "Coronavirus Disease 2019 (COVID-19)." Centers for Disease Control and Prevention, February 11, 2020. Updated May 20, 2020. https://www.cdc.gov/coronavirus/2019-ncov/ hcp/planning-scenarios.html. Archived: https://www.cdc.gov/

coronavirus/2019-ncov/hcp/planning-scenarios-archive/
planning-scenarios-2020-05-20.pdf.

[29] Chalmers, Vanessa. "Two-Metre Rule Has NO Basis in
Science, Leading Scientists Say." Mail Online, June 16, 2020.
https://www.dailymail.co.uk/news/article-8425671/Two-metre-
rule-NO-basis-science-leading-scientists-say-amid-calls-drop-
measure.html.

[30] Xiao, J., Shiu, E., Gao, H., Wong, J. Y., Fong, M. W., Ryu,
S....Cowling, B. J. (May 2020). Nonpharmaceutical Measures for
Pandemic Influenza in Nonhealthcare Settings—Personal
Protective and Environmental Measures. Emerging Infectious
Diseases, 26(5), 967-975. https://dx.doi.org/10.3201/
eid2605.190994.

[31] Jacobs, J. L., Ohde, S., Takahashi, O., Tokuda, Y., Omata, F.,
& Fukui, T. (June 2009). Use of surgical face masks to reduce the
incidence of the common cold among health care workers in
Japan: a randomized controlled trial. American journal of
infection control, 37(5), 417–419. https://doi.org/10.1016/j.ajic.
2008.11.002.

[32] Cowling, B. J., Y. Zhou, D. K. M. Ip, G. M. Leung, and A. E.
Aiello. "Face Masks to Prevent Transmission of Influenza Virus:
A Systematic Review." Epidemiology & Infection 138, no. 4
(January 22, 2010): 449–56. https://doi.org/10.1017/
S0950268809991658.

[33] bin−Reza, Faisal, Vicente Lopez Chavarrias, Angus Nicoll,
and Mary E. Chamberland. "The Use of Masks and Respirators
to Prevent Transmission of Influenza: A Systematic Review of
the Scientific Evidence." Influenza and Other Respiratory
Viruses 6, no. 4 (December 21, 2011): 257–67. https://doi.org/
10.1111/j.1750-2659.2011.00307.x.

[34] Smith, Jeffrey D., Colin C. MacDougall, Jennie Johnstone,
Ray A. Copes, Brian Schwartz, and Gary E. Garber.

"Effectiveness of N95 Respirators versus Surgical Masks in Protecting Health Care Workers from Acute Respiratory Infection: A Systematic Review and Meta-Analysis." CMAJ 188, no. 8 (May 17, 2016): 567–74. https://doi.org/10.1503/cmaj.150835.

[35] Offeddu, Vittoria, Chee Fu Yung, Mabel Sheau Fong Low, and Clarence C. Tam. "Effectiveness of Masks and Respirators Against Respiratory Infections in Healthcare Workers: A Systematic Review and Meta-Analysis." Clinical Infectious Diseases 65, no. 11 (November 13, 2017): 1934–42. https://doi.org/10.1093/cid/cix681.

[36] Radonovich, Lewis J., Michael S. Simberkoff, Mary T. Bessesen, Alexandria C. Brown, Derek A. T. Cummings, Charlotte A. Gaydos, Jenna G. Los, et al. "N95 Respirators vs Medical Masks for Preventing Influenza Among Health Care Personnel: A Randomized Clinical Trial." JAMA 322, no. 9 (September 3, 2019): 824–33. https://doi.org/10.1001/jama.2019.11645.

[37] Long, Youlin, Tengyue Hu, Liqin Liu, Rui Chen, Qiong Guo, Liu Yang, Yifan Cheng, Jin Huang, and Liang Du. "Effectiveness of N95 Respirators versus Surgical Masks against Influenza: A Systematic Review and Meta-Analysis." Journal of Evidence-Based Medicine 13, no. 2 (2020): 93–101. https://doi.org/10.1111/jebm.12381.

[38] Blitzer, Wolf. "CNN.Com - Search for the 'smoking Gun' - Jan. 10, 2003." https://www.cnn.com/2003/US/01/10/wbr.smoking.gun/.

[39] Iacurci, Greg. "Unemployment Is Nearing Great Depression Levels. Here's How the Eras Are Similar — and Different." CNBC, May 19, 2020. https://www.cnbc.com/2020/05/19/unemployment-today-vs-the-great-depression-how-do-the-eras-compare.html.

[40] Van Dam, Andrew, Long, Heather. "U.S. Unemployment Rate Soars to 14.7 Percent, the Worst since the Depression Era." Washington Post. May 8, 2020. https://www.washingtonpost.com/business/2020/05/08/april-2020-jobs-report/.

[41] Kochhar, Rakesh. "Unemployment Rose Higher in Three Months of COVID-19 than It Did in Two Years of the Great Recession." Pew Research Center (blog). June 11, 2020. https://www.pewresearch.org/fact-tank/2020/06/11/unemployment-rose-higher-in-three-months-of-covid-19-than-it-did-in-two-years-of-the-great-recession/.

[42] International Institute of Forecasters. "Forecasting for COVID-19 Has Failed," June 14, 2020. https://forecasters.org/blog/2020/06/14/forecasting-for-covid-19-has-failed/.

[43] STAT. "In the Coronavirus Pandemic, We're Making Decisions without Reliable Data," March 17, 2020. https://www.statnews.com/2020/03/17/a-fiasco-in-the-making-as-the-coronavirus-pandemic-takes-hold-we-are-making-decisions-without-reliable-data/.

[44] Miltimore, Jon. "Modelers Were 'Astronomically Wrong' in COVID-19 Predictions, Says Leading Epidemiologist—and the World Is Paying the Price | Jon Miltimore," July 2, 2020. https://fee.org/articles/modelers-were-astronomically-wrong-in-covid-19-predictions-says-leading-epidemiologist-and-the-world-is-paying-the-price/.

[45] Kesslen, Ben. "Dr. Birx Predicts up to 200,000 U.S. Coronavirus Deaths 'If We Do Things Almost Perfectly.'" CNBC, March 30, 2020. https://www.cnbc.com/2020/03/30/white-house-coronavirus-expert-predicts-up-to-200000-us-coronavirus-deaths.html.

[46] Healthcare Purchasing News. "COVID-19 Predicted to Infect 81% of U.S. Population, Cause 2.2 Million Deaths in U.S.," March 18, 2020. https://www.hpnonline.com/infection-

prevention/screening-surveillance/article/21130206/covid19-predicted-to-infect-81-of-us-population-cause-22-million-deaths-in-us.

[47] Rushton, Katherine, and Daniel Foggo. "Neil Ferguson, the Scientist Who Convinced Boris Johnson of UK Coronavirus Lockdown, Criticised in Past for Flawed Research." The Telegraph, March 28, 2020. https://www.telegraph.co.uk/news/2020/03/28/neil-ferguson-scientist-convinced-boris-johnson-uk-coronavirus-lockdown-criticised/.

[48] Mcgraw, Meridith. "Trump's New Coronavirus Argument: 2 Million People Are Being Saved." POLITICO. Accessed July 15, 2020. https://www.politico.com/news/2020/04/01/trump-coronavirus-millions-saved-160814.

[49] AP NEWS. "Many Failures Combined to Unleash Death on Italy's Lombardy," April 26, 2020. https://apnews.com/de2794327607a3a67ed551f0b6b71404.

[50] "Hospital 'a Week Away' from Being Overrun." BBC News, May 15, 2020, sec. Wales. https://www.bbc.com/news/uk-wales-52673516.

[51] Colarossi, Natalie. "36 Photos Show How New York Is Getting through the World's Biggest Coronavirus Outbreak." Business Insider. Accessed July 27, 2020. https://www.businessinsider.com/photos-show-how-new-york-is-grappling-with-coronavirus-outbreak-2020-3.

[52] Miles, Frank. "NYC Hospitals 'overwhelmed' by Coronavirus Patients, Resident Warns." Text.Article. Fox News. Fox News, March 26, 2020. https://www.foxnews.com/health/nyc-hospitals-overwhelmed-by-coronavirus-patients-resident-warns.

[53] NPR.org. "U.S. Field Hospitals Stand Down, Most Without Treating Any COVID-19 Patients." Accessed July 27, 2020. https://www.npr.org/2020/05/07/851712311/u-s-field-

hospitals-stand-down-most-without-treating-any-covid-19-patients.

[54] Gonen, Yoav. "Brooklyn Field Hospital Shuts After $21 Million Construction — and Zero Patients." THE CITY, May 21, 2020. https://www.thecity.nyc/health/2020/5/21/21273179/brooklyn-field-hospital-shuts-after-21-million-construction-and-zero-patients.

[55] "The Coronavirus Field Hospitals That Weren't." Accessed July 27, 2020. https://www.ny1.com/nyc/all-boroughs/news/2020/04/23/coronavirus-field-hospitals-that-weren-t.

[56] Admin, LBBJ. "Hospitals Weren't Overrun by COVID-19, but Now They Face a New Challenge: The Budget." Long Beach Business Journal (blog), May 5, 2020. https://www.lbbusinessjournal.com/hospitals-werent-overrun-by-covid-19-but-now-they-face-a-new-challenge-the-budget/.

[57] "COVID-19 Empty Hospitals in Videos by Citizen Reporters." Accessed July 27, 2020. https://153news.net/watch_video.php?v=87WORR9O73UY.

[58] "WORLD'S BIGGEST LIE EMPTY HOSPITALS." Accessed July 27, 2020. https://153news.net/watch_video.php?v=N7G64K7DM1U6.

[59] NBC News. "Coronavirus Deniers Take Aim at Hospitals as Pandemic Grows." Accessed July 27, 2020. https://www.nbcnews.com/tech/social-media/coronavirus-deniers-take-aim-hospitals-pandemic-grows-n1172336.

[60] COVID-19 Alert No. 2, CDC (United States Centers for Disease Control) website, March 24, 2020. Steven Schwartz, PhD Director – Division of Vital Statistics. National Center for Health Statistics 3311 Toledo Rd | Hyattsville, MD 20782. https://www.cdc.gov/nchs/data/nvss/coronavirus/Alert-2-New-ICD-code-introduced-for-COVID-19-deaths.pdf.

[61] "New ICD Code Introduced for COVID-19 Deaths," March 24, 2020, 1.
https://www.who.int/classifications/icd/covid19/en/
https://www.who.int/classifications/icd/COVID-19-coding-icd10.pdf.

[62] Dr. Anthony Fauci, director of the National Institute of Allergy and Infectious Diseases, National Institutes of health (U.S.), speaking at a Whitehouse Coronavirus Taskforce Briefing on 7 April 2020: https://youtu.be/MJFx3ZC3FR8?t=9338.

[63] Dr. Deborah Birx, Coronavirus Response Coordinator for the White House Coronavirus Task Force, speaking at a Whitehouse Coronavirus Taskforce Briefing on 7 April 2020: https://youtu.be/MJFx3ZC3FR8?t=9272.

[64] Schuman, T., & Prophet, E. C. (1984). Soviet ideological subversion of America in four stages: Elizabeth Clare Prophet interviews Tomas Schuman, Novosti Press, Soviet defector. Malibu, CA: Summit University. http://archive.org/details/IdeologicalSubversion.

[65] World Health Organization. (2020). Laboratory testing for coronavirus disease 2019 (COVID-19) in suspected human cases: interim guidance, 2 March 2020. World Health Organization, p. 4. https://apps.who.int/iris/handle/10665/331329.

[66] Gralinski, Lisa E., and Vineet D. Menachery. "Return of the Coronavirus: 2019-NCoV." Viruses 12, no. 2 (January 24, 2020). https://doi.org/10.3390/v12020135.

[67] "The Pathogenicity of SARS-CoV-2 in HACE2 Transgenic Mice | BioRxiv." Accessed July 5, 2020. https://www.biorxiv.org/content/10.1101/2020.02.07.939389v3.

[68] Wu, Fan, Su Zhao, Bin Yu, Yan-Mei Chen, Wen Wang, Zhi-Gang Song, Yi Hu, et al. "A New Coronavirus Associated with Human Respiratory Disease in China." Nature 579, no. 7798

(March 2020): 265–69. https://doi.org/10.1038/
s41586-020-2008-3.

[69] Zhou, Peng, Xing-Lou Yang, Xian-Guang Wang, Ben Hu,
Lei Zhang, Wei Zhang, Hao-Rui Si, et al. "Discovery of a Novel
Coronavirus Associated with the Recent Pneumonia Outbreak in
Humans and Its Potential Bat Origin." BioRxiv, January 23, 2020,
2020.01.22.914952. https://doi.org/10.1101/2020.01.22.914952.

[70] Huang, Chaolin, Yeming Wang, Xingwang Li, Lili Ren,
Jianping Zhao, Yi Hu, Li Zhang, et al. "Clinical Features of
Patients Infected with 2019 Novel Coronavirus in Wuhan,
China." The Lancet 395, no. 10223 (February 15, 2020): 497–506.
https://doi.org/10.1016/S0140-6736(20)30183-5.

[71] Bao, Linlin, Wei Deng, Baoying Huang, Hong Gao, Jiangning
Liu, Lili Ren, Qiang Wei, et al. "The Pathogenicity of SARS-
CoV-2 in HACE2 Transgenic Mice." BioRxiv, February 28, 2020,
2020.02.07.939389. https://doi.org/10.1101/2020.02.07.939389.

[72] Peiris, JSM, ST Lai, LLM Poon, Y Guan, LYC Yam, W Lim,
J Nicholls, et al. "Coronavirus as a Possible Cause of Severe
Acute Respiratory Syndrome." The Lancet 361, no. 9366 (April
19, 2003): 1319–25. https://doi.org/10.1016/
S0140-6736(03)13077-2.

[73] Fouchier, Ron A. M., Thijs Kuiken, Martin Schutten, Geert
van Amerongen, Gerard J. J. van Doornum, Bernadette G. van
den Hoogen, Malik Peiris, Wilina Lim, Klaus Stöhr, and Albert D.
M. E. Osterhaus. "Koch's Postulates Fulfilled for SARS Virus."
Nature 423, no. 6937 (May 2003): 240–240. https://doi.org/
10.1038/423240a.

[74] Zhu, Na, Dingyu Zhang, Wenling Wang, Xingwang Li, Bo
Yang, Jingdong Song, Xiang Zhao, et al. "A Novel Coronavirus
from Patients with Pneumonia in China, 2019." New England
Journal of Medicine 382, no. 8 (February 20, 2020): 727–33.
https://doi.org/10.1056/NEJMoa2001017.

[75] Kim, Jeong-Min, Yoon-Seok Chung, Hye Jun Jo, Nam-Joo Lee, Mi Seon Kim, Sang Hee Woo, Sehee Park, Jee Woong Kim, Heui Man Kim, and Myung-Guk Han. "Identification of Coronavirus Isolated from a Patient in Korea with COVID-19." Osong Public Health and Research Perspectives 11, no. 1 (February 2020): 3–7. https://doi.org/10.24171/j.phrp. 2020.11.1.02.

[76] Mossman, Karen. "I Study Viruses: How Our Team Isolated the New Coronavirus to Fight the Global Pandemic." McMaster University (March 25, 2020). https://brighterworld.mcmaster.ca/ articles/i-study-viruses-how-our-team-isolated-the-new-coronavirus-to-fight-the-global-pandemic/.

[77] "Robert Koch." In Wikipedia, July 29, 2020. https:// en.wikipedia.org/w/index.php? title=Robert_Koch&oldid=970161449.

[78] Koch, R. (1876). "Untersuchungen über Bakterien: V. Die Ätiologie der Milzbrand-Krankheit, begründet auf die Entwicklungsgeschichte des Bacillus anthracis" [Investigations into bacteria: V. The etiology of anthrax, based on the ontogenesis of Bacillus anthracis]. Cohns Beitrage zur Biologie der Pflanzen (in German). 2 (2): 277–310.

[79] Isolating a virus mainly involves three steps: 1) filtration by particle size, 2) centrifugation (filtration by particle density) and 3) morphological identification via electron microscopy. A plain English description of the process can be found here: https:// www.andrewkaufmanmd.com/the-rooster-in-the-river-of-rats/ and is mirrored by several other video hosts; for example: https://youtu.be/Z4Za6uNAFvM, https://www.bitchute.com/ video/Zuy88IdwsFxt/ and https://www.bitchute.com/video/ THJw9B5EMvkk/.

[81] Andrew Kaufmann, MD. Website: https:// www.andrewkaufmanmd.com/episodes-medicamentum-authentica/. YouTube channel: https://www.youtube.com/ channel/UCV7v2cvSnrJ9Qyz36cW1Ftw/videos.

[82] Devereux, Amory, Frei, Rosemary. "Scientists Have Utterly Failed to Prove That the Coronavirus Fulfills Koch's Postulates," OffGuardian, June 9, 2020. https://off-guardian.org/2020/06/09/scientists-have-utterly-failed-to-prove-that-the-coronavirus-fulfills-kochs-postulates/.

[83] "Dr. Kary Banks Mullis." Accessed September 15, 2020. http://www.karymullis.com/pcr.shtml.

[84] NobelPrize.org. "The Nobel Prize in Chemistry 1993." https://www.nobelprize.org/prizes/chemistry/1993/mullis/lecture/.

[85] ("Quantitative PCR is an oxymoron.") "Kary Mullis - AIDS Wiki." http://aidswiki.net/index.php/Kary_Mullis.

[86] CDC. "Information for Laboratories about Coronavirus (COVID-19)." Centers for Disease Control and Prevention, February 11, 2020. https://www.cdc.gov/coronavirus/2019-ncov/lab/grows-virus-cell-culture.html.

[87] Harcourt, Jennifer, Azaibi Tamin, Xiaoyan Lu, Shifaq Kamili, Senthil K. Sakthivel, Janna Murray, Krista Queen, et al. "Severe Acute Respiratory Syndrome Coronavirus 2 from Patient with Coronavirus Disease, United States - Volume 26, Number 6—June 2020 - Emerging Infectious Diseases Journal - CDC." Accessed July 17, 2020. https://doi.org/10.3201/eid2606.200516.

[88] Berry, M., A. Gurung, and D. L. Easty. "Toxicity of Antibiotics and Antifungals on Cultured Human Corneal Cells: Effect of Mixing, Exposure and Concentration." Eye 9, no. 1 (January 1995): 110–15. https://doi.org/10.1038/eye.1995.17.

[89] PromoCell. "Antibiotics in Cell Culture: Friend or Enemy?," January 29, 2018. https://www.promocell.com/in-the-lab/antibiotics-in-cell-culture-friend-or-enemy/.

[90] Németh, A., Orgovan, N., Sódar, B.W. et al. Antibiotic-induced release of small extracellular vesicles (exosomes) with surface-associated DNA. Sci Rep 7, 8202 (2017). https://doi.org/10.1038/s41598-017-08392-1.

[91] The World Health Organization. "Coronavirus disease 2019 (COVID-19)
Situation Report – 46", March 6, 2020. https://www.who.int/docs/default-source/coronaviruse/situation-reports/20200306-sitrep-46-covid-19.pdf?sfvrsn=96b04adf_4.

[92] RevCycleIntelligence. "Hospital Reimbursement for Uninsured COVID-19 Cases May Total $42B." RevCycleIntelligence, April 8, 2020. https://revcycleintelligence.com/news/hospital-reimbursement-for-uninsured-covid-19-cases-may-total-42b.

[93] "DNA Vaccination." In Wikipedia, July 25, 2020. https://en.wikipedia.org/w/index.php?title=DNA_vaccination&oldid=969383431.

[94] Research, Center for Drug Evaluation and. "Coronavirus Treatment Acceleration Program (CTAP)." FDA, July 14, 2020. https://www.fda.gov/drugs/coronavirus-covid-19-drugs/coronavirus-treatment-acceleration-program-ctap.

[95] D'Souza, Gypsyamber, Dowdy, David. Johns Hopkins, Bloomberg School of Public Health. "What Is Herd Immunity and How Can We Achieve It With COVID-19?" Johns Hopkins Bloomberg School of Public Health. APRIL 10, 2020. https://www.jhsph.edu/covid-19/articles/achieving-herd-immunity-with-covid19.html.

[96] "Herd Immunity (Herd Protection) | Vaccine Knowledge." Accessed July 8, 2020. https://vk.ovg.ox.ac.uk/vk/herd-immunity.

[97] Fine, Paul, Ken Eames, and David L. Heymann. "'Herd Immunity': A Rough Guide." Clinical Infectious Diseases 52, no. 7 (April 1, 2011): 911–16. https://doi.org/10.1093/cid/cir007.

[98] Gordis, L. (2013). Epidemiology E-Book. United Kingdom: Elsevier Health Sciences. pp. 26-27.

[99] Jacobson v. Massachusetts, 197 U.S. 11 (1905).

[100] Buck v. Bell, 274 U.S. 200 (1927).

[101] Garden State Observer. "Dershowitz Says Government 'Police Power' to Force Coronavirus Vaccinations Is 'Not Debatable,'" May 21, 2020. https://gardenstateobserver.com/dershowitz-says-government-police-power-to-force-coronavirus-vaccinations-is-not-debatable/.

[102] "Allopathic Medicine." In Wikipedia, July 10, 2020. https://en.wikipedia.org/w/index.php?title=Allopathic_medicine&oldid=966988120.

[103] National Constitution Center – constitutioncenter.org. "The Mysterious Death of George Washington - National Constitution Center." December 14, 2019. https://constitutioncenter.org/blog/the-mysterious-death-of-george-washington.

[104] Pedersen, N., Kang, L. (2017). Quackery: A Brief History of the Worst Ways to Cure Everything. United States: Workman Publishing Company.

[105] "Saddam Hussein and Al-Qaeda Link Allegations." In Wikipedia, July 3, 2020. https://en.wikipedia.org/w/index.php?title=Saddam_Hussein_and_al-Qaeda_link_allegations&oldid=965875629.

[106] Abc-Clio Information Services., Lawson, R. M. (2004). Science in the Ancient World: An Encyclopedia. United Kingdom: ABC-CLIO.

[107] "Geocentric Model." In Wikipedia, June 30, 2020. https://en.wikipedia.org/w/index.php?title=Geocentric_model&oldid=965229665.

[108] Centers for Disease Control (CDC). Measles outbreak among vaccinated high school students--Illinois. MMWR Morb Mortal Wkly Rep. 1984;33(24):349-351.

[109] Nkowane, B M et al. "Measles outbreak in a vaccinated school population: epidemiology, chains of transmission and the role of vaccine failures." American journal of public health vol. 77,4 (1987): 434-8. doi:10.2105/ajph.77.4.434.

[110] Rosen, Jennifer B., Jennifer S. Rota, Carole J. Hickman, Sun B. Sowers, Sara Mercader, Paul A. Rota, William J. Bellini, et al. "Outbreak of Measles Among Persons With Prior Evidence of Immunity, New York City, 2011." Clinical Infectious Diseases 58, no. 9 (May 1, 2014): 1205–10. https://doi.org/10.1093/cid/ciu105.

[111] Hersh, Bradley S., Paul E. M. Fine, W. Kay Kent, Stephen L. Cochi, Laura H. Kahn, Elizabeth R. Zell, Patrick L. Hays, and Cindy L. Wood. "Mumps Outbreak in a Highly Vaccinated Population." The Journal of Pediatrics 119, no. 2 (August 1, 1991): 187–93. https://doi.org/10.1016/S0022-3476(05)80726-7.

[112] Rubin, Steven A., Malen A. Link, Christian J. Sauder, Cheryl Zhang, Laurie Ngo, Bert K. Rima, and W. Paul Duprex. "Recent Mumps Outbreaks in Vaccinated Populations: No Evidence of Immune Escape." Journal of Virology 86, no. 1 (January 1, 2012): 615–20. https://doi.org/10.1128/JVI.06125-11.

[113] Patel, Leena N., Robert J. Arciuolo, Jie Fu, Francesca R. Giancotti, Jane R. Zucker, Jennifer L. Rakeman, and Jennifer B. Rosen. "Mumps Outbreak Among a Highly Vaccinated University Community—New York City, January–April 2014." Clinical Infectious Diseases 64, no. 4 (February 15, 2017): 408–12. https://doi.org/10.1093/cid/ciw762.

[114] Dachel, A., Kennedy, R. F. (2017). Vaccine Villains: What the American Public Should Know about the Industry. United States: Skyhorse Publishing Company, Incorporated.

[115] Ioannidis, John P A. "Why most published research findings are false." PLoS medicine vol. 2,8 (2005): e124. doi: 10.1371/journal.pmed.0020124.

[116] Steinhoff, Mark C., Joanne Katz, Janet A. Englund, Subarna K. Khatry, Laxman Shrestha, Jane Kuypers, Laveta Stewart, et al. "Year-Round Influenza Immunisation during Pregnancy in Nepal: A Phase 4, Randomised, Placebo-Controlled Trial." The Lancet. Infectious Diseases 17, no. 9 (2017): 981–89. https://doi.org/10.1016/S1473-3099(17)30252-9.

[117] James M. McCarty, Michael D. Lock, Kristin M. Hunt, Jakub K. Simon, Marc Gurwith, Safety and immunogenicity of single-dose live oral cholera vaccine strain CVD 103-HgR in healthy adults age 18–45, Vaccine, Vol. 36, Issue 6, 2018, pp. 833-840, ISSN 0264-410X, https://doi.org/10.1016/j.vaccine.2017.12.062.

[118] Reisinger EC, Tschismarov R, Beubler E, et al. Immunogenicity, safety, and tolerability of the measles-vectored chikungunya virus vaccine MV-CHIK: a double-blind, randomised, placebo-controlled and active-controlled phase 2 trial. Lancet. 2019;392(10165):2718-2727. doi:10.1016/S0140-6736(18)32488-7.

[119] Marshall HS, Richmond PC, Beeslaar J, et al. Meningococcal serogroup B-specific responses after vaccination with bivalent rLP2086: 4 year follow-up of a randomised, single-blind, placebo-controlled, phase 2 trial. Lancet Infect Dis. 2017;17(1):58-67. doi:10.1016/S1473-3099(16)30314-0.

[120] Protein Sciences Corporation. "Evaluation of the Immunogenicity, Safety, Reactogenicity, Efficacy, Effectiveness and Lot Consistency of FluBlok® Trivalent Recombinant Baculovirus-Expressed Hemagglutinin Influenza Vaccine In

Healthy Adults Aged 18 to 49." Clinical trial registration. clinicaltrials.gov, May 16, 2011. https://clinicaltrials.gov/ct2/show/NCT00539981.

[121] Vaccines. "Polio - Vaccines - ProCon.Org." Accessed July 24, 2020. https://vaccines.procon.org/vaccine-histories-and-impact/polio/.

[122] CDC. "Polio Elimination in the U.S." Centers for Disease Control and Prevention, October 25, 2019. https://www.cdc.gov/polio/what-is-polio/polio-us.html.

[123] Iannelli, Vincent, and MD. "When Was the Last Measles Death in the United States?" VAXOPEDIA, April 15, 2018. https://vaxopedia.org/2018/04/15/when-was-the-last-measles-death-in-the-united-states/.

[124] "The Case AGAINST Vaccination By M. BEDDOW BAYLY M.R.C.S., L.R.C.P." Accessed July 24, 2020. http://www.whale.to/vaccines/bayly.html.

[125] Guyer, B., M. A. Freedman, D. M. Strobino, and E. J. Sondik. "Annual Summary of Vital Statistics: Trends in the Health of Americans during the 20th Century." Pediatrics 106, no. 6 (December 2000): 1307–17. https://doi.org/10.1542/peds.106.6.1307.

[126] Waxman, Henry A. "H.R.5546 - 99th Congress (1985-1986): National Childhood Vaccine Injury Act of 1986." Webpage, October 18, 1986. 1985/1986. https://www.congress.gov/bill/99th-congress/house-bill/5546.

[127] Stern, W. (1912). Die psychologischen Methoden der Intelligenzprüfung und deren Anwendung an Schulkindern .... Germany: J.A. Barth.

[128] Sternberg, R. J. (2018). Theories of intelligence. In S. I. Pfeiffer, E. Shaunessy-Dedrick, & M. Foley-Nicpon (Eds.), APA

handbooks in psychology®. APA handbook of giftedness and talent (p. 145–161). American Psychological Association.

[129] Buck v. Bell, 274 US 200 (Supreme Court 292AD). https://scholar.google.com/scholar_case?q=Buck+v.+Bell.+274+U.S.+200+(1927).&hl=en&as_sdt=806&case=1700304772805702914&scilh=0.

[130] Yan, X.-H., Boyer, T., Trenberth, K., Karl, T.R., Xie, S.-P., Nieves, V., Tung, K.-K. and Roemmich, D. (2016), The global warming hiatus: Slowdown or redistribution?. Earth's Future, 4: 472-482. doi:10.1002/2016EF000417 https://agupubs.onlinelibrary.wiley.com/doi/full/10.1002/2016EF000417.

[131] Spearman, C. (1923). The Nature of "intelligence" and the Principles of Cognition. United Kingdom: Macmillan.

[132a] Whipple, G. M., Stern, W. (1914). The Psychological Methods of Testing Intelligence. United States: Warwick & York.

[132b] Kaufman, Alan S.; Lichtenberger, Elizabeth (2006). Assessing Adolescent and Adult Intelligence (3rd ed.). Hoboken (NJ): Wiley.

[132c] Saccuzzo, D. P., Kaplan, R. M. (2009). Psychological Testing: Principles, Applications, and Issues. United States: Cengage Learning.

[132d] Das, J. P., Kirby, J. R., & Jarman R. F. (1975). Simultaneous and successive syntheses: An alternative model for cognitive abilities. Psychological Bulletin, 82, 87–103.

[132e] Daniel, Mark H. (1995). "Differential Ability Scales". In Sternberg, Robert J. (ed.). Encyclopedia of human intelligence. 1. Macmillan. pp. 350–354.

[133]
Galbraith, J., Bartholomew, D. J., Steele, F., Moustaki, I. (2008). A nalysis of Multivariate Social Science Data. United States: CRC Press.

[134] Jensen, A. R. (1998). The *g* Factor: The Science of Mental Ability. United Kingdom: Praeger.

[135] Spearman, C. (2005). The Abilities of Man: Their Nature and Measurement. United States: Blackburn Press.

[136] Jolliffe, I. (2013). Principal Component Analysis. United States: Springer New York.

[137] Cohen, J., Pfeiffer, K. & Francis, J.A. Warm Arctic episodes linked with increased frequency of extreme winter weather in the United States. Nat Commun 9, 869 (2018). https://doi.org/ 10.1038/s41467-018-02992-9.

[138] Thomas R. Karl et. al. "Possible artifacts of data biases in the recent global surface warming hiatus" Science. 26 Jun 2015: Vol. 348, Issue 6242, pp. 1469-1472. DOI: 10.1126/ science.aaa5632. https://science.sciencemag.org/content/ 348/6242/1469.abstract? ijkey=.l.kxQb89CJjY&keytype=ref&siteid=sci.

[139] "Climate-Change 'Hiatus' Disappears with New Data: Nature News & Comment." 4 June 2015. https:// www.nature.com/news/climate-change-hiatus-disappears-with- new-data-1.17700.

[140] Cowtan, K. and Way, R.G. (2014), Coverage bias in the HadCRUT4 temperature series and its impact on recent temperature trends. Q.J.R. Meteorol. Soc., 140: 1935-1944. doi: 10.1002/qj.2297.

[141] Schmidt, G., Shindell, D. & Tsigaridis, K. Reconciling warming trends. Nature Geosci 7, 158–160 (2014). https:// doi.org/10.1038/ngeo2105.

[142] Vaccine insert for RECOMBIVAX HB® (hepatitis b). December 2018 (first approved 1986). Sec. 6.1 Clinical Trials Experience, p. 4. https://www.fda.gov/files/vaccines%2C %20blood%20%26%20biologics/published/package-insert-recombivax-hb.pdf.

[143] Ernie Tretkoff. "This Month in Physics History." American Physical Society. June 2006: Vol. 15, No. 6. http://www.aps.org/publications/apsnews/200606/history.cfm.

[144] World Health Assembly, 72. "WHO Results Report: Programme Budget 2018-2019: Mid-Term Review," 2019. https://apps.who.int/iris/handle/10665/328787.

[145] "CrossFit Settles Lawsuit With HHS After Agency Releases Emails Showing Continued Efforts to Conceal Donations." Accessed August 2, 2020. https://www.crossfit.com/battles/crossfit-settles-lawsuit-with-hhs-after-agency-agrees-to-release-redacted-emails.

[146] Says, Mary W. "Congress Presses for Transparency at Groups Supporting NIH, CDC." STAT (blog), July 2, 2018. https://www.statnews.com/2018/07/02/congress-transparency-funding-nih-cdc/.

[147] MIT Technology Review. "This Is What It Will Take to Get Us Back Outside." Accessed August 2, 2020. https://www.technologyreview.com/2020/04/12/999117/blueprint-what-it-will-take-to-live-in-a-world-with-covid-19/.

[148] Gallagher, James. "When Will the Coronavirus Outbreak End?" BBC News, March 23, 2020, sec. Health. https://www.bbc.com/news/health-51963486.

[149] Renwick, Danielle. "How Quickly Will There Be a Vaccine? And What If People Refuse to Get It?" The Guardian, July 16, 2020, sec. US news. https://www.theguardian.com/us-news/2020/jul/16/coronavirus-vaccine-covid-19-experts.

[150] Tsipursky, Gleb. "Bad News about the Pandemic: We're Not Getting Back to Normal Any Time Soon." Scientific American. Accessed August 2, 2020. https://www.scientificamerican.com/article/bad-news-about-the-pandemic-were-not-getting-back-to-normal-any-time-soon/.

[151] "Student's t-Distribution." In Wikipedia, July 25, 2020. https://en.wikipedia.org/w/index.php?title=Student%27s_t-distribution&oldid=969494195.

[152] "Scientific Method." In Wikipedia, August 3, 2020. https://en.wikipedia.org/w/index.php?title=Scientific_method&oldid=970998954.

[153] National Geographic News. "Chimps, Humans 96 Percent the Same, Gene Study Finds," August 31, 2005. https://www.nationalgeographic.com/news/2005/8/chimps-humans-96-percent-the-same-gene-study-finds/.

[154a] "COVID-19 Provisional Counts - Weekly Updates by Select Demographic and Geographic Characteristics." Accessed September 18, 2020. https://www.cdc.gov/nchs/nvss/vsrr/covid_weekly/index.htm.

[154b] Bondy, Dave. "CDC: 94% of Covid-19 Deaths Had Underlying Medical Conditions." MSN. September 1, 2020. Accessed September 18, 2020. https://www.msn.com/en-us/Health/medical/cdc-94-25-of-covid-19-deaths-had-underlying-medical-conditions/ar-BB18wrA7

[155] Newey, Sarah. "Why Have so Many Coronavirus Patients Died in Italy?" The Telegraph, March 19, 2020. https://www.telegraph.co.uk/global-health/science-and-disease/have-many-coronavirus-patients-died-italy/.

[156] Elizabeth Cohen and Dana Vigue, CNN Health. "US Government Slow to Act as Anti-Vaxxers Spread Lies on Social Media about Coronavirus Vaccine." CNN. August 12, 2020.

https://www.cnn.com/2020/08/12/health/anti-vaxxers-covid-19/index.html.

[157] Official web site of the U.S. Health Resources & Services Administration. "About the National Vaccine Injury Compensation Program." Text, May 11, 2017. https://www.hrsa.gov/vaccine-compensation/about/index.html.

[158] "Publication 510 (02/2020), Excise Taxes | Internal Revenue Service." Accessed September 9, 2020. https://www.irs.gov/publications/p510.

[159] "Vaccine Adverse Event Reporting System (VAERS)." Accessed September 9, 2020. https://vaers.hhs.gov/.

[160] Moseman, E. Ashley, Matteo Iannacone, Lidia Bosurgi, Elena Tonti, Nicolas Chevrier, Alexei Tumanov, Yang-Xin Fu, Nir Hacohen, and Ulrich H. von Andrian. "B Cell Maintenance of Subcapsular Sinus Macrophages Protects against a Fatal Viral Infection Independent of Adaptive Immunity." Immunity 36, no. 3 (March 23, 2012): 415–26. https://doi.org/10.1016/j.immuni.2012.01.013. https://www.sciencedirect.com/science/article/pii/S107476131200057X.

[161] Balko, Radley. "It Literally Started with a Witch Hunt: A History of Bite Mark Evidence." Washington Post. Accessed September 21, 2020. https://www.washingtonpost.com/news/the-watch/wp/2015/02/17/it-literally-started-with-a-witch-hunt-a-history-of-bite-mark-evidence/.

[162] Salem Witch Museum. "Spectral Evidence," February 15, 2013. https://salemwitchmuseum.com/2013/02/15/spectral-evidence/.

www.ingramcontent.com/pod-product-compliance
Lightning Source LLC
Chambersburg PA
CBHW072141090426
42739CB00013B/3245

* 9 7 8 1 7 3 5 7 5 6 8 3 7 *